The Dama Gazelles

Publication assisted by the SECOND ARK FOUNDATION
working with the EXOTIC WILDLIFE ASSOCIATION

Number 58
W. L. Moody Jr. Natural History Series

The

Above left: Eastern dama gazelle, familiar in Texas (also called addra or red-necked gazelle) (photo by Elizabeth Cary Mungall, courtesy of Kyle Wildlife, Texas, USA).

Center: Central dama gazelle (rarest of all, and none known outside Africa) (photo by Bertrand Chardonnet of animal held at Douguia on the shore of Lake Chad).

Right: Western dama gazelle (known as mhorr gazelle or mohor) (photo by Abdelkader Jebali, courtesy of North Ferlo Fauna Reserve, Senegal).

Texas A&M University Press
College Station

Dama Gazelles

Last Members of a Critically Endangered Species

Edited by Elizabeth Cary Mungall

With Special Contributions by

Teresa Abáigar
Lisa Banfield with Hessa Al Qahtani and Mark Craig
Adam Eyres
Tania Gilbert with Gerardo Espeso Pajares
Abdelkader Jebali
Andrew C. Kitchener
John Newby with Tim Wacher and Thomas Rabeil
Helen Senn
Frans M. van den Brink

And a Foreword by Bonnie C. Yates and an Introduction by David Mallon

First edition

This paper meets the requirements of ANSI/NISO Z39.48-1992 (Permanence of Paper).
Binding materials have been chosen for durability.
Manufactured in China through FCI Print Group

Library of Congress Cataloging-in-Publication Data

Names: Mungall, Elizabeth Cary, author.
Title: The dama gazelles: last members of a critically endangered species / Elizabeth Cary Mungall; with special contributions by Teresa Abáigar [and 15 others].
Other titles: W.L. Moody Jr. natural history series; no. 58.
Description: First edition. | College Station: Texas A&M University Press, [2018] | Series: W. L. Moody Jr. natural history series; number 58 | Includes bibliographical references and index.
Identifiers: LCCN 2017035144 | ISBN 9781623496111 (printed case: alk. paper)
Subjects: LCSH: Gazelles. | Gazelles—Reintroduction. | Endangered species.
Classification: LCC QL737.U53 .M84 2018 | DDC 599.64/69—dc23 LC record available at https://lccn.loc.gov/2017035144

To the late Wesley W. Kyle, who worked long and hard to establish a herd of the eastern dama gazelle on his Texas ranch, and to his daughter, Kathryn Kyle, and her husband, Scott A. Smith, who grew the herd with a conservation program—and allowed me in for long-term research.

Wesley W. Kyle (photo by Christian Mungall, courtesy of Kyle Wildlife, Texas, USA).

Mar Cano (photo by T. Abáigar, Spain).

To the late Mar Cano, who worked with Professor José Antonio Valverde and her father, Antonio Cano, on the rescue operation that saved the western dama gazelle, also called mhorr gazelle, from extinction and then dedicated her life to the conservation of this gazelle.

And to all those who work for the survival of the dama gazelles.

Contents

Foreword

The impact of humans on the environment is evident in the case of dama gazelles. Although the gazelles were once commonly found in their natural African habitat, wild numbers have dwindled to dangerously low levels. The majority of the world's dama gazelles now live on Texas ranches, with additional individuals in zoos and preserves around the world. As Elizabeth Cary Mungall observes in this volume, the situation for the dama gazelle is "precarious, but not hopeless." This volume offers a critical wealth of information meant to help rebuild the species.

Authoritative and expansive in scope, this multidisciplinary approach will be eminently valuable to conservation organizations, ranchers, and regulators. Given the important role Texas landowners play in the story of the dama gazelle's chances for survival, I especially appreciate the range of the contributors' expertise as well as the volume's acknowledgment of the achievements and additions to science by private facilities, not just by traditional zoos. Also important is the book's explanation of the new Source Population Alliance. Its potential to coordinate joint conservation efforts among ranches, zoos, and range states is an important development designed to help the species survive. The encyclopedic variety of topics in this book lends itself to the interests of an equally varied population of readers.

The species history, including the history of captive populations in zoos and similar institutions, reads like the "genealogy" of a species—facts that are usually deeply buried in hard-to-find documents or stored only in the memories of people who were there. The chapter by Frans van den Brink, who first brought individuals of this species out of Africa—individuals from the eastern part of their range—provides an enlightening view in a first-person account. I applaud its inclusion in the book. Coupled with the rescue details of how Professor José Antonio Valverde acted on his realization that this species was about to go extinct in the western part of its range, this demonstrates how critical the actions of a few committed people can be to species survival.

As for the multitude of information in both text and pictures, the most valuable to me is the precise morphological details of coat color and change, manifestations of sexual dimorphism, and variation in horn shape of different ages and sexes. Such information cannot be found in field guides dealing with many species in a particular geographic locality. Here it is, at least for this species. Having spent much of my working career with the US Fish and Wildlife Service, I know firsthand how crucial these sorts of details are to the accurate identification needed by wildlife forensic scientists charged with distinguishing hides, horns, and bones of endangered animals.

Population statistics are included and are augmented by additional observations about transport survival, changes through time, and physical descriptions of the natural range sites other than just size. Explanation of aspects such as the most important vegetation, territory size limitations, and ideas (e.g., adding brush breaks) and ways to segregate breeding groups will make this book indispensable.

Finally, the chapter on hand raising gazelles will make the book very popular with anyone responsible for successfully rearing a young gazelle. I find the writing here to be humorous but full of good lessons and handy hints, including helpful warnings. The glossary is an added bonus because not all readers will be biologists. With such broad coverage in the book's chapters, the reference section becomes a compendium of treatises on the species.

I find this book to be an ideal source of information about conservation methodology for the species. The pertinent data gleaned from experienced handlers and biologists from so many of the wild locales and captive dama gazelle programs in the world will be a great resource for range states working to ensure survival of this species. Baseline statistics on physical growth and population dynamics, reproductive details, and subsistence needs are all helpful, if not vital, to herd managers in order to identify problems in time to address them.

In closing, I would like to convey my congratulations to the authors for providing a work that underscores how serious human encroachment is, whether by agricultural practices or by vegetation removal. It is seriously detrimental to the survival of this species of wild gazelle as well as to the people in their environment. I can deduce that Mungall's purpose for compiling this information is ultimately to help "build back the wildlife heritage of Africa."

Bonnie C. Yates

Acknowledgments

As editor of this book on dama gazelles, I send my heartfelt thanks to all the authors who joined me in making this a truly comprehensive overview. You have all helped stress the conservation needs, initiatives, and possibilities for this species on the brink of extinction. Can the dama gazelle be saved? I know that you all hope so, and that this has been a major motivation behind the hours and effort you have given for this species, including taking time out from other important duties to share your expertise with the readers of this book. There are also other committed people out there working hard for the same goals, but the following are investigators whom I was able to meet at the 2013 dama gazelle conservation workshop hosted by the Royal Zoological Society of Scotland and who, along with a few others recruited along the way, responded to this book project as one more way to seek assistance for the dama gazelle. In alphabetical order by first chapter author, with coauthors grouped together, joining me in the writing were Teresa Abáigar; Lisa Banfield, Hessa Al Qahtani, and Mark Craig; Adam Eyres; Tania Gilbert and Gerardo Espeso Pajares; Abdelkader Jebali; Andrew C. Kitchener; David Mallon; John Newby, Tim Wacher, and Thomas Rabeil; Helen Senn; and Frans van den Brink. Many of you have submitted illustrations as well as text. For this I am grateful, too. How can a book about a beautiful animal catch readers' attention without pictures?

At this point, I would also like to thank all the people and organizations who have made it possible for me to delve into the life habits of the dama gazelle. Without you, I would never have been able to assemble so many topics, much less write about a number of them myself. Among you are Mark C. Reed and his director, Louis DiSabato, at the San Antonio Zoo. You introduced me to the dama gazelle at a time when individuals imported by Frans van den Brink were still in the collection. I was privileged to be able to observe the dama gazelle, a species that always intrigued my major professor, Fritz R. Walther at Texas A&M University, although he never had the opportunity to study it himself. Then came all my friends at Kyle Wildlife. You established a dama gazelle herd as a conservation project and you have humored me so many times, whether it be getting weights, teeth, and DNA samples; keeping notes on a hand-raised fawn; granting permission to run a series of 24-hour studies at designated phases of the moon; allowing me a bed to sleep on before and after dawn-to-dark watches; providing meals that were so very good and so very welcome, and on and on. There were Wes and Jackie Kyle, who started the herd that I have been watching for so long; Kathryn Kyle and Scott A. Smith, who have been keeping the herd going; Brigitta Richardson, who did translation work and also let me know what she had seen on the ranch; and Pablo Cervantes, Shawn Rowland, James "Sarge" Stanick, Mark Valvo, and Corey Villa, who also filled me in on what the gazelles had been doing when I was not there.

A recent addition to the Kyle Wildlife work has been GPS-radio collaring in pastures of different sizes and vegetative characteristics. Steve Forest of Stevens Forest Ranch in West Texas even mounted a helicopter capture operation to get collars on a selection of his dama gazelle males. Carlos Chappa III was the expert marksman as Steve Forest flew. Susan M. Cooper and her team of Shane Sieckenius and Andrea Silva from Texas A&M AgriLife Research and Extension Center in Uvalde, Texas, did an expert job fitting the collars and then were instrumental in the tracking and data analysis. For the comparison that Kevin and Cole Reid allowed at their Morani River Ranch (now organized as Stewards of Wildlife Conservation) in the Texas Hill Country, Susan and her team were on hand again. John Fredericks and the rest of the staff worked with Cole Reid on capture and handling and secured the animals while blood was drawn for DNA analysis. Kenneth Haynie and Loyce "Jim" Deggs added extra insights along the way.

The Exotic Wildlife Association (EWA), the Second Ark Foundation (SAF), and their members have aided at so many critical times with access, information, and monetary support. EWA executive director Charly Seale, EWA president Don Tarpey, and his wife, SAF president Bobbie Tarpey, have been instrumental in arranging what was necessary for various aspects of the long-term dama gazelle project to go forward. Larry Johnson of Safari Enterprises let me make a series of visits to his mhorr gazelle herd. Douglas E. Smith of Bear Creek Exotics contributed records from dama gazelles that he had hand raised. Tim Fallon of the Texas Hill Country Chapter of Safari Club International arranged funding to start the collaring investigations, and Kevin Reid of Morani River Ranch put in the final amount needed for collaring to begin.

I am indebted to all the organizations that have provided support. In addition to EWA and SAF, there have been Bisbee's Fish and Wildlife Conservation Fund (Wayne Bisbee and Brian White), the Fossil Rim Wildlife Center (Pat R. Condy), the Texas Hill Country Chapter and the Austin Chapter of Safari Club International (Tim Fallon), the Dallas Safari Club (Ben Carter), and the Greater Houston Chapter of the American Association of Zoo Keepers.

In addition, my students at Texas Woman's University, Department of Biology, have helped with lively comments when dama gazelle examples have turned up in class, and Lynda Uphouse and members of the Institutional Animal Care and Use Committee have put through my "use" request for dama gazelle hand raising, even though the stated means for lowering stress during handling for measurements of height and chest girth was offering potato chips instead of administering drugs.

Throughout all this, there has been my husband and best companion, Christian Mungall, who has allowed himself to be pulled in as photographer, data analyst, idea man, and all the other "hats" that have taken him away from doing what he might rather have been involved with at the time.

This book is possible only because of all of you.

Elizabeth Cary Mungall
Kerrville, Texas
June 2017

Introduction

The tall and elegant dama gazelle is one of the world's rarest antelopes (figure below). Its distinctive chestnut and white coat sets it apart from all other gazelles. The amounts of chestnut or reddish hair and white hair vary in the eastern and western forms, known as "addra" (or "red-necked gazelle") and "mhorr gazelle" (or "mohor gazelle"), respectively. Captive populations of both these forms are managed separately and, together with another, have been categorized as subspecies. Recent research, however, shows that genetic differences among the three forms are not highly marked.

Figure. Bachelor male of the eastern dama gazelle, known as "addra," in display to another male (photo by Elizabeth Cary Mungall, courtesy of Kyle Wildlife, Texas, USA).

The dama gazelle once ranged from the Atlantic almost to the River Nile in the semideserts and dry Sahel grasslands fringing the Sahara to the north and south. Its decline began a long time ago. Details of the former range north of the Sahara are not known, but it is believed to have disappeared from that part of its range by the end of the nineteenth century, or very early in the twentieth century.

The current distribution of the dama gazelle is limited to five small fragments—two in Niger, two in Chad, and one in Mali—although there has been no information on the last of these subpopulations since 2007. Together, these represent a tiny fraction, less than 1 percent, of the original range.

The numbers remaining in the wild may be 100 to 200, and it is even possible that there are fewer than 100 adults of breeding age left. The dama gazelle is perilously close to vanishing completely from the wild.

The cause of this huge reduction in distribution and numbers is the all-too-familiar combination of overhunting, overgrazing, habitat loss, increasing temperatures, and the expansion of desert conditions. Contributing factors include the drilling of boreholes for water that allow herders and their livestock to remain in an area year round, improved roads that facilitate access by poachers and cause disturbance, and widespread instability and insecurity that prevent effective patrolling and other conservation activities.

Captive dama gazelles held in zoos and similar collections around the world number 500 to 600. Most of these are managed under cooperative breeding programs organized by the Association of Zoos and Aquariums (AZA) in North America and the European Association of Zoos and Aquaria (EAZA) in Europe.

Texas ranches hold more than 1,000 dama gazelles—more than those in the wild and in zoos put together. Some of these facilities have habitats and climatic conditions similar to those in the natural range in North Africa.

The captive and semicaptive populations of dama gazelles are all of great importance in safeguarding the species for the future, for producing stock for reintroductions, and for research into behavior, foraging, and home-range size. Animals on display are also important as "species ambassadors" to show to the public, hardly any of whom will ever have the opportunity to see a dama gazelle in the wild.

Several attempts have been made to reintroduce dama gazelles back into the wild in Morocco, Senegal, and Tunisia. All these projects have started with captive-born stock released into fenced enclosures within national parks and reserves. In May 2015, animals from one of these reserves in Western Sahara were even set free. Results are being monitored. All these programs have been relatively small scale, so they are not yet regarded as successful reintroductions. Although suitable habitat still remains in some places, full reintroduction into a wild state is not feasible in any part of the region for the time being, because of the threat of poaching. It is difficult to predict when the situation will change for the better.

It is also of concern that some attempts at release into fenced areas have succeeded, while others have not. These mixed results show that the dama gazelle is a sensitive species. Reasons have not yet been established, but dama gazelles appear to have specific requirements regarding diet, habitat, enclosure size, herd structure, and presence of other species. Research into the factors resulting in the success or failure of establishing herds is clearly needed.

An outline conservation strategy for the dama gazelle was developed at a dama gazelle workshop held in Edinburgh, Scotland, in November 2013, followed by consultation with many additional experts on the species. The strategy included a shared vision: "Sustainable and free-living populations of dama gazelles in their native range, supported by well managed populations elsewhere."

Attaining this vision and safeguarding the long-term future of the species will require a joint effort by all those involved and in all four settings where dama gazelles are found—in the wild, in countries of their former range where they have been repatriated, in zoos and similar collections, and on Texas and other US ranches. All have an important role to play in research and conservation.

This book describes the history, taxonomy, genetics, biology, behavior, and wild status of the dama gazelle for a wider audience, summarizing its current plight and mobilizing support for its survival over generations to come.

David Mallon

The Dama Gazelles

Section 1 The Dama Gazelle through Time

This book is divided into four sections. Each emphasizes a different focus of the ongoing work with dama gazelles. Section 1 asks what the dama gazelles are and why they are vanishing. Chapter 1 gives a historical overview of the disappearance of the dama gazelles from so much of their native range. Although accelerated by modern developments in firearms and all-terrain vehicles, this decline has been going on for a long time.

Chapter 2 shows that attempts to define the maximum original distribution of dama gazelles use correlations with favored types of habitat as well as with locations of ancient artwork. These supplement the fragmentary record. Environmental changes over geologic time take the discussion into an examination of dama gazelle taxonomy. Comparing past possibilities with what can be documented from modern times portrays a once wide-ranging species that has been cut into small, isolated populations. A situation that was once more fluid has become a northwest-to-southeast series of varied populations that have on average less and less of the reddish coat color and more and more of the white. To recognize these shifting averages and to facilitate discussions about the different segments of the whole, this continuum—or "cline"—has been designated as a series of three "subspecies." For dama gazelles, "subspecies" is a semantic category that struggles to make discrete segments for today's situation out of a coordinated whole that has been changing over time. Because of how separated the few remaining dama gazelle populations have become, a subspecies model seems reasonable.

As discussed in chapter 3 on genetics, scientists doing genetic testing—at least of DNA passed down from mothers to offspring as opposed to the contribution from fathers—have found no correlation of inherited groups of genes with the different color patterns characteristic of different parts of the range, but even these scientists find the traditional terminology of three subspecies helpful "for ease of reference" (RZSS and IUCN Antelope Specialist Group 2014). Certainly, short of human intervention, nothing like the former connectivity is ever going to be reestablished. But why all this worry over taxonomy and genetics? Not only taxonomists in a theoretical sense but also zoo staff charged with sustaining their particular herds of animals and conservationists considering reintroductions must wrestle with the question of which animals are appropriate to use and which ones should be put together. Maintaining as much genetic diversity as possible is a big challenge when there are only small numbers of animals available as mates—only small numbers of gene possibilities (alleles) that can be passed on. The array of gene forms gives groups flexibility to stay attuned to their surroundings, especially to changes.

At least for now, the remaining wild populations each show a varied gene complement, although this is different for each. In contrast, none of the captive populations in Europe, on the Arabian Peninsula, or even in African breeding centers shows more than one, or at most two, different versions for its inherited gene groups, and there are only three versions among them all. Only the flourishing herds of the eastern dama gazelle, or addra, in the United States come close. They show five possibilities. Both the larger numbers of breeding animals and the habit of managers drawing on multiple sources to form new herds, as well as switching out the breeding male periodically so that fathers will not be forever breeding their own offspring, help retain more of the greater genetic diversity that the United States is assumed to have started with from the original import. All of this reduces the chances of animals inheriting the same versions of their genes from both parents and, thus, reduces inbreeding. Sometimes, a double dose of the same allele works fine, but other times inheriting the same kind of allele from both parents has harmful effects. All these aspects deserve attention in order to maximize long-term chances for species survival.

With so very little diversity left in the captive herds of the western dama gazelle—called mhorr gazelle, an animal that went extinct in the wild—conservationists have started carefully controlled experiments to see what happens under zoo conditions when some of these western dama gazelles are allowed to breed with eastern dama gazelles like those common in Texas. The first-generation offspring lose the distinctive color pattern of their parents, but, presumably, they now have two instead of one kind of inherited gene cluster. Perhaps this little increase in genetic diversity can help prolong the species even if it contributes to a second extinction event for the mhorr gazelle. If this sort of interbreeding comes to the United States, it could further the genetic diversity of the crossbreeds, but it could also lead to the disappearance of the eastern dama gazelle, a subspecies that currently has the numbers (as of 2015, 1,510 on ranches—chiefly in Texas—in addition to approximately 150 in US zoos) as well as the management to continue on its own.

Seeing a respected group of scientists breeding the different subspecies together, Texas ranchers may be likely to start their own experimentation. If the species became managed as a whole rather than as three parts, then owners would see no reason not to let any dama gazelle in with any other dama gazelle. Eventually, the result would be the same as what has happened with fallow deer in Texas, sika deer in the United States, and animals like zebras and giraffes and others the world over—in captivity, the species would continue, but no longer with its distinctive, purebred parts. And crossbreeds could get released to bolster diminishing wild populations. Pros and cons of different scenarios will continue to be discussed, but the ultimate goal is for the dama gazelle to go on.

Chapter 4 finishes Section 1 with a historical series of artworks. They illustrate how dama gazelles have been depicted in successive ages and indicate the changing uses of dama gazelle art as chronicles through time. Each century has had artists who have taken dama gazelles as their subjects. This helps show us the many facets of dama gazelles.

Chapter 1

Path to Endangered Status

Abdelkader Jebali

From a species formerly common across a broad band of dry steppe and Sahelian savanna country on the fringe of Africa's Sahara, with extensions into the desert, the dama gazelle has continued to shrink in distribution. This decrease has continued right up to the present time. Wars as well as modern vehicles and weapons have led to excessive hunting in increasingly remote areas just as habitat loss from drought, domestic livestock overgrazing, and more permanent settlements of pastoralists have pushed wildlife farther into marginal habitat.

Only five small, free-ranging populations are known in addition to the May 2015 release into the wild of 24 mhorr gazelles. (For a discussion of the five, see chapter 20. For details of the release, see chapter 19.) The decline of the dama gazelle is part of the near-total collapse of the large Sahelo-Saharan fauna (Jebali 2008, Durant et al. 2014). Prominent species such as the addax (*Addax nasomaculatus*), scimitar-horned oryx (*Oryx dammah*), red-necked ostrich (*Struthio camelus camelus*), and Saharan cheetah (*Acinonyx jubatus hecki*) that used to share part or all of the same ecosystems with the dama gazelle vanished in less than one century.

Figure 1.1 shows an example of the kind of animal that will be lost if the dama gazelle goes extinct. With supporting maps, this chapter shows the pattern of the dama gazelle's decline from the mid-nineteenth century to the early twenty-first. Causes are also discussed.

Figure 1.1. Mhorr gazelle buck passes a group of scimitar-horned oryx, a species that went extinct in the wild (photo by Abdelkader Jebali, courtesy of North Ferlo Fauna Reserve, Senegal).

Distribution and Pattern of Decline

In 1850, the dama gazelle was widespread among 13 North African countries (map 1.1). However, it was very rare in Tunisia (Pervinquière 1912) and had probably vanished before the twentieth century (Kacem et al. 1994b). In Egypt, the species was unknown, and data on its presence in the historical period are lacking (Flower 1932, Osborn and Helmy 1980, Manlius 1996, Saleh 2001).

In the Sahelian zone south of the Sahara, dama gazelles used to occur in high densities, and undoubtedly populations continued to be numerous until the first half of the twentieth century. On the other hand, this gazelle seems always to have been rare in some Saharan areas such as around the Tibesti Massif and some ergs and in heavily wooded southern Sahel areas. Maps 1.1 and 1.2 show that the distribution of the dama gazelle remained practically the same until the beginning of the twentieth century. Nevertheless, an early decline was noted locally in areas where modern weapons became increasingly numerous following African colonization (In Tanoust 1930).

This decline first began in the southern provinces of Morocco (Lavauden 1926), where mhorr gazelle numbers (numbers of the western subspecies of dama gazelle, *Nanger dama mhorr*) showed some fluctuations before starting to break down in the 1950s (map 1.3). In the Sahel, the species remained relatively common and even locally abundant in countries such as Mali, Niger, Chad, and Sudan (Maydon 1923, Lavauden 1926, Bourbon 1929, Joleaud 1929, Borricand 1945, Lhote 1946, Audas 1951, Wilson 1980). Roughly near the border between Niger and Chad, these countries see the central dama gazelle (*Nanger dama dama*) grade into the eastern subspecies, addra (*Nanger dama ruficollis*). Reduction was mainly at the edges of the range, but it also followed the expanded movement of troops.

The decade from 1940 to 1950 inaugurated an irreversible process in the decline of the dama gazelle. The population decrease accelerated after 1950, occasioning the fragmentation of the gazelle range across the Sahelo-Saharan region (map 1.4). In this "atomization" of the overall population, Chad

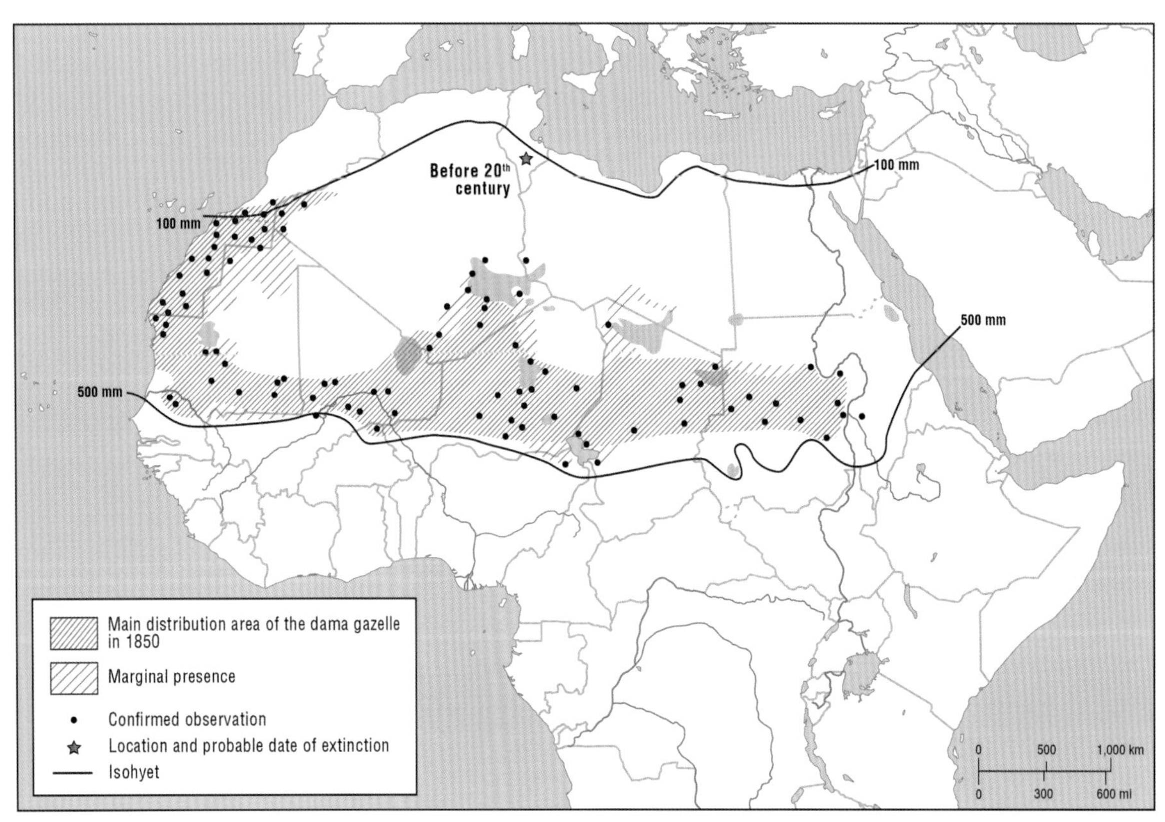

Map 1.1. Distributional area of dama gazelles in 1850 (probable extinction in southern Tunisia sometime between the seventeenth and nineteenth centuries).

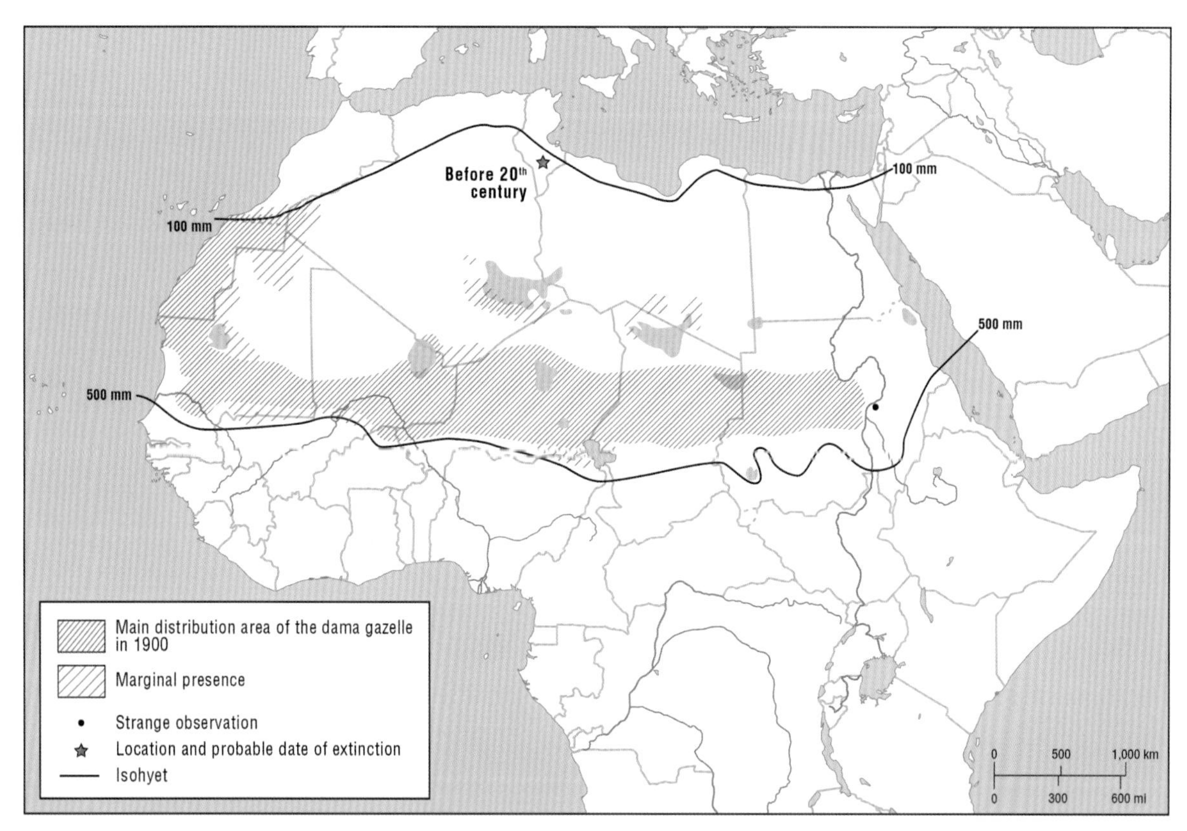

Map 1.2. Distributional area of dama gazelles in 1900.

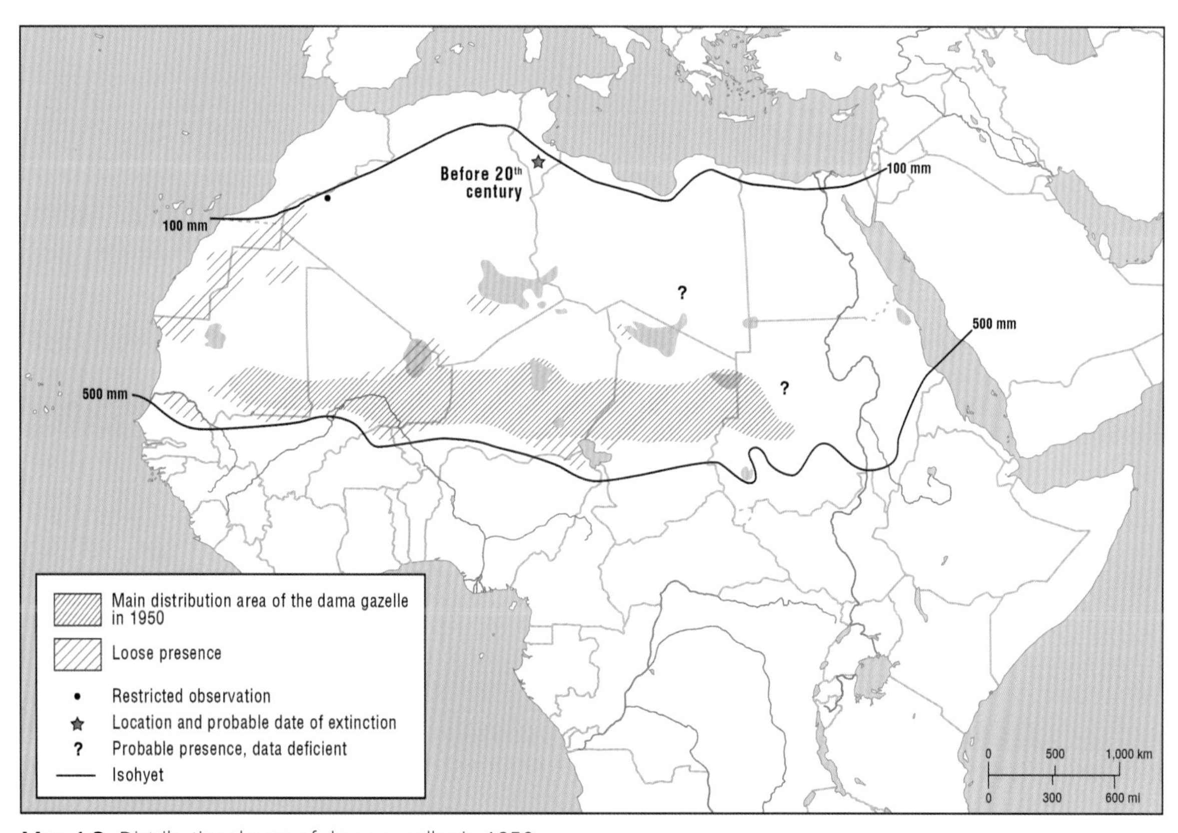

Map 1.3. Distributional area of dama gazelles in 1950.

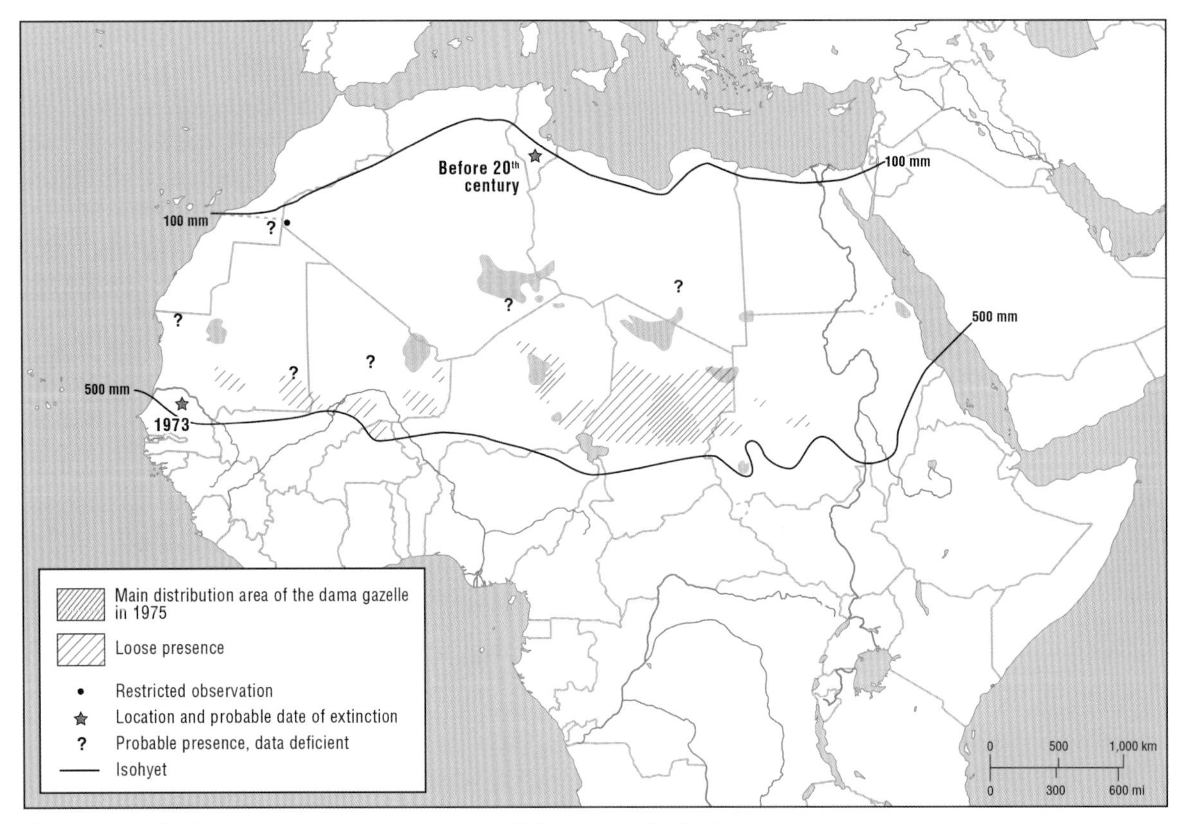

Map 1.4. Distributional area of dama gazelles in 1975 (mhorr gazelle finally disappeared from Western Sahara in 1968 and from Senegal in 1973).

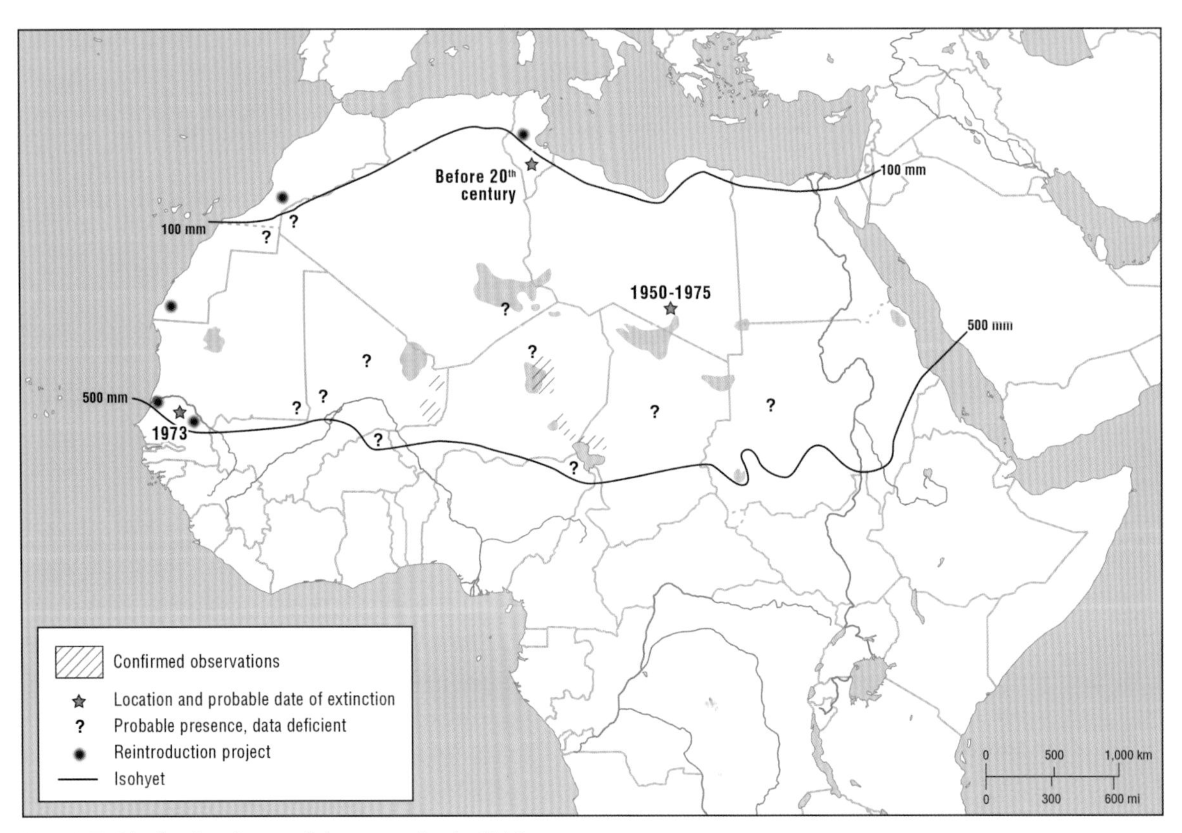

Map 1.5. Distributional area of dama gazelles in 2010.

remained home to the largest concentration of dama gazelles. In the other countries, the rapid decline of the species ended in local extinction (map 1.4), as in Senegal, where the dama gazelle became extinct in 1973 at the latest, after being seen for the last time in the early 1970s in the Fété Olé region of northern Ferlo (Poulet 1972, 1974). In the Mauritanian Atlantic Sahara, the last individuals were recorded in the late 1960s in and around the current Banc d'Arguin National Park (Trotignon 1975, Dubreuil 1987). In 1968, the mhorr gazelle disappeared from Western Sahara, the area of the capture discussed in chapter 17 that has allowed captive breeding to keep this subspecies alive (Cano 1991). Elsewhere, the presence of the species was reduced to a few relict groups scattered over only a small portion of its former range (Dupuy 1967, Sayer 1977, Wilson 1980, Loggers et al. 1992).

Between 1975 and 2005, the dama gazelle's distribution shrank alarmingly. Although in the mid-1970s there were considerable numbers of wild dama gazelles in their Chad stronghold, military action from 1978 to 1987 brought marked reductions (East 1990). Map 1.5 illustrates the serious geographic regression in 2010 compared with the previous situation. Since 2000, the occurrence of wild dama gazelles has been confirmed in only 3 countries of the 13 they inhabited in the nineteenth century: the neighboring countries of Chad, Niger, and Mali. Such a decline is often accompanied by a major "crash" in numbers, as is often the case for animal populations under heavy pressure from fragmentation and hunting. In Aïr and Ténéré National Nature Reserve, attrition of the dama gazelle population between 1982 and 1990 was confirmed by the decreasing number of observations as well as by the increasingly smaller size of herds (Poilecot 1996a). Moreover, relict subpopulations were increasingly spread over vast areas. Wacher et al. (2004) mentioned that in a survey that involved the Manga region of Chad in 2001, only 15 dama gazelles were sighted as observation progressed over a 250 km (155 mi) route. Surprisingly, in the Ouadi Rimé–Ouadi Achim Game Reserve, a major site for the species in the mid-1970s, when there were an estimated 6,000 to 8,000 dama gazelles (East 1990), no dama gazelles were seen at all (Newby 1981, Grettenberger and Newby 1986, Thomassey and Newby 1990, Emanoil and IUCN 1994).

Elsewhere, apart from Mali and Niger, where small relict populations have been reported (Kriska 2001, Claro 2004, Lamarque and Niagaté 2004, Wacher et al. 2004, Lamarque 2005) over very long distances, extinction has been confirmed for some countries (Senegal, Tunisia), and an "unknown" status has been attributed to others because this gazelle has not been seen for so long (Burkina Faso, Libya, Mauritania, Nigeria, Sudan) or because of the uncertainties of unreliable data (Algeria, Morocco). Researchers are monitoring the May 2015 experimental release of mhorr gazelles into the wild in Safia Nature Reserve to see whether this western subspecies of dama gazelle can reestablish itself in southern Morocco.

Discussion and Conclusion

From a widespread and abundant species in the nineteenth century, the dama gazelle has shrunk to tiny, localized populations. The present survey shows that dama gazelle distribution declined from the 1950s onwards, and there were local decreases in the preceding decades. As part of this trend, range fragmentation, starting in the 1940s to 1950s, has become irreversible. In 1988, the International Union for Conservation of Nature (IUCN) listed the dama gazelle as an endangered species and attributed to it the status of a vulnerable taxon (VU). Before this, the overall status of the species was not well known even though its decline on both the eastern and western edges of its range was already noticeable. The weakening of the dama gazelle populations accelerated when the last concentrations of the species, namely in Mali, Niger, and Chad, were mostly diminished. In less than 20 years, the dama gazelle became a critically endangered (CR) species.

The drop in dama gazelle numbers was mostly the result of two major factors: excessive killing and expansion of pastoralism (Newby 1980, Heringa 1990, Heringa et al. 1990, Thomassey and Newby 1990, Emanoil and IUCN 1994, Poilecot 1996a). These two phenomena were sometimes combined with other human or natural factors, such as rebellions or drought, driving the dama gazelle to its present state of scarcity:

- Excessive killing (hunting and poaching). There is almost unanimous agreement that excessive killing is the main cause of the decline of the dama gazelle (Heim de Balsac 1958, Newby 1980, Vincke 1987, Heringa 1990, Heringa et al. 1990, Cloudsley-Thompson 1992, Emanoil and IUCN 1994, Poilecot 1996a, Kriska 2001, Cuzin 2003). An alarming reduction has been noted since the 1950s, especially following political and social unrest.

- Expansion of pastoralism. Since the 1950s, the Sahel has undergone great changes because of the introduction of deep, cemented wells and boreholes (Newby 1982). The presence of water throughout the year has encouraged a permanent presence of nomadic pastoralists who have broadened their scope (Heringa 1990). Previously, pastures at these latitudes were reached only during the rainy season. The gradual settlement has brought with it overgrazing and degradation of habitat (Emanoil and IUCN 1994).

Today, survival of the species hangs in the balance since fewer than 300 individuals are thought still to be living in the wild (RZSS and IUCN Antelope Specialist Group 2014). Any resurgence will certainly involve only a portion of the previous distribution. Questions of inbreeding, genetic diversity, and subspecies divisions are matters that will gain increasing importance with such fragmented populations living at such low numbers.

Chapter 2

Taxonomy and Distribution

What Is a Dama Gazelle?

Andrew C. Kitchener

All mammalogists agree that the dama gazelle is a "good species," an animal that everyone agrees is a particular kind with no confusion with other kinds. What is not clear is its geographical variation and the number of subspecies that should be recognized (Cano Perez 1984, Drüwa 1985, Groves and Grubb 2011). Sadly, today the dama gazelle survives in only a few very small, fragmented populations, and museum specimens of known origin are scarce (Cano Perez 1984, RZSS and IUCN Antelope Specialist Group 2014). With little hard evidence to go on, it is difficult to piece together unequivocal evidence for the original geographical variation of the dama gazelle. Its taxonomic history has been very dynamic, with subdivision into as many as three species and several subspecies (Cano Perez 1984, Groves and Grubb 2011). However, the successful conservation of the dama gazelle depends on a robust understanding of its intraspecific taxonomy, so it is important to review current evidence to help make informed decisions about its conservation both in the wild and in captivity.

How Many Subspecies?

All species vary, whether at an individual level, between the sexes, or geographically. Geographical variation may be taxonomically important and indicate that a population has evolved more or less in isolation, with characteristics that distinguish it from most individuals from other populations of the same species. However, there is no precise definition of a subspecies that is applied universally, although often a 75 percent rule is used. According to this rule, 75 percent of individuals from one population must be distinguishable from 100 percent of individuals from another population if the two populations are to be considered different subspecies (Mayr and Ashlock 1991). Therefore, the problem comes in drawing the dividing line between apparently different populations and in knowing whether apparent differences are of any taxonomic significance. This situation can be exacerbated by limited numbers of specimens from the original range, which may allow only a partial, or even false, picture of former geographical variation.

Today, the dama gazelle is usually recognized as a single species that is subdivided into three subspecies, which, from east to west and northwest, have increasing amounts of red or reddish-brown coloration on the neck, back, flanks, and legs (Cano Perez 1984, Groves and Grubb 2011). Some mammalogists have suggested that this represents a cline showing a gradual change in coloration across the Sahel of North Africa, but as we shall see, the situation is more complex than this. Also, it is inappropriate to apply subspecies names to a cline, because there is significant gene flow among the populations and, thus, subspecies names become arbitrary labels for portions of the cline. It is possible that this may have happened to the dama gazelle.

The three dama gazelle subspecies that are currently recognized across North Africa are the following (fig. 2.1):

Nanger dama mhorr in the northwest (mhorr gazelle)

Nanger dama dama in the west and center (dama gazelle)

Nanger dama ruficollis in the east (addra or red-necked gazelle)

Figure 2.1. Examples of the three principal color forms of dama gazelles—northwestern, central, and eastern: *A*, mhorr gazelle descended from stock caught in Western Sahara (photo by A. C. Kitchener, courtesy of Frankfurt Zoo, Germany, copyright © A. C. Kitchener); *B*, dama gazelle from Termit in the central part of the distribution, Niger (camera-trap photo from the wild, courtesy of Thomas Rabeil, copyright © Thomas Rabeil/Sahara Conservation Fund); *C*, addra descended from stock caught in northeastern Chad show examples of color variation common among addra in this part of their distribution (adult and adolescent males, *left*, and three adult females, *right*) (photo by Elizabeth Cary Mungall, courtesy of Second Ark Foundation GPS-radio collaring project at Morani River Ranch, Texas, USA).

A

B

C

Given that some gene flow may occur between subspecies, the use of this taxonomic rank to describe apparently more or less discrete morphological forms, which occur in different contiguous geographical regions, would seem reasonable. However, a recent molecular study of mitochondrial DNA was unable to find any genetic structure that correlated with the three putative subspecies (Senn et al. 2014; see chapter 3). This study has been updated with further specimens from North American captive collections and a wild population from the Aïr Mountains in Niger (Senn et al. 2016), which confirm this overall picture of a lack of genetic structure across the range of the species. Obviously, there is some genetic basis for the observed morphological variation, because mhorr and addra gazelles breed true in captivity, although there may be considerable individual variation within the same herd (fig. 2.1*C*; see discussion below and in chapter 8). Further genetic studies, including those on nuclear DNA and the Y chromosome, are required to see whether this lack of genetic structure is confirmed among dama gazelles. However, this chapter provides details of experimental crossbreeding of addra and mhorr gazelles at Al Ain Zoo (Anonymous 2016).

Taxonomy has changed considerably over the last 250 years since Linnaeus developed the system for scientific names that we all use today. From brief physical descriptions in the eighteenth and early nineteenth centuries, through detailed, but still largely discursive, descriptions in the late nineteenth and early twentieth centuries, to the potential for detailed molecular and morphological studies today, the tools of the taxonomic trade have been constantly changing as new technologies and techniques have been developed. There have also been significant recent changes in how we define species, in particular from the Biological Species Concept (BSC) to the Phylogenetic Species Concept (PSC). This has led to an enormous increase in the number of recognized species today as many subspecies are raised to species level (e.g. Groves and Grubb 2011). Because of the lack of breeding between putative species as defined by the BSC, the bar has been considerably lowered to populations having at least one unique or diagnosable character that allows them to be distinguished, mostly visually, under the PSC. Certainly, the dama gazelle has not been immune to taxonomic changes since its formal description by Peter Pallas in 1766.

Evolving Taxonomy

The dama gazelle was once included in the genus *Gazella*, which formerly included all gazelle species (e.g. Corbet and Hill 1991). However, the dama gazelle was later placed in the genus *Nanger* (Lataste 1885), although this was not widely recognized. The genus *Nanger* is restricted to the three largest gazelles: the species complex comprising the dama gazelle, Soemmerring's gazelle (*N. soemmerringii*), and Grant's gazelle (*N. granti*). Some researchers have shown that these three largest gazelles all group together in molecular phylogenies but are not associated with the other gazelles; hence they must be recognized as belonging to a distinct genus (Bärmann et al. 2013).

As well as cranial characters, the three *Nanger* gazelles share a distinctive rump patch, unlike that of other gazelles. (For a detailed discussion of the *Nanger* rump patch, see chapter 8.) As fawns grow, the darker dorsal coloration fades to white at the rump to form prongs that extend forward on each side of the body, while the white of the rump patch joins with the white spreading onto the tail and croup. (For a diagram showing the names of gazelle body parts, such as "croup," see appendix 2.)

When Linnaeus (1758) described antelopes, including gazelles, using his novel system, he placed them all in the single genus *Capra*, which is now restricted to goats and ibexes. Peter Pallas (1766) placed antelopes, including the dama gazelle, in the new genus *Antilope* and hence first described it as *Antilope dama*. As mammalogists were able to establish more and more morphological characters to distinguish between different antelopes, so the number of genera increased. Therefore, the genus *Antilope* is now reserved for the species that first received this name—the blackbuck, *Antilope cervicapra*, of India—while gazelles were placed in the genus *Gazella*, proposed by Blainville (1816). Interestingly, Blainville placed *Antilope dama* in the genus *Cervicapra*, now a junior synonym of *Antilope*. Recent molecular phylogenies have shown that gazelles do not form a cohesive genetic group. They have been split into three genera, *Gazella*, *Nanger*, and *Eudorcas* (Bärmann et al. 2013) (table 2.1).

Table 2.1. Dama gazelle scientific names past and present, with geographic indications

SPECIES	SUBSPECIES	OTHER PROPOSALS
Original name *Antilope dama* by Peter S. Pallas 1766 From Senegal		
Familiar designation *Gazella dama* changed by John Edward Gray 1846	Familiar designations Three subspecies by Mar Cano Perez 1984 *Gazella dama dama* Cano Perez – from western and southern part of distribution	*Gazella dama permista* by Oscar Neumann 1906 From Senegal *Gazella dama damergouensis* by Walter Rothschild 1921 From Takoukout, Damergou, Niger River
	Gazella dama ruficollis Originally *Antilope ruficollis* by Charles Hamilton Smith 1827 From Dongola, Sudan Cano Perez – from eastern and toward central part of distribution	
	Gazella dama mhorr Originally *Antilope mhorr* by Edward Bennett 1833 From Wednun, near Tafilat, Mogador, Morocco Cano Perez – from northwestern part of distribution	*Gazella dama lozanoi* by Eugenio Morales Agacino 1934 From Cape Juby, Río de Oro, Western Sahara
New designation *Nanger dama* By F. Lataste 1885, but not confirmed until 1999 by W. Rebholz and E. Harley	New designations *Nanger dama dama* *Nanger dama ruficollis* *Nanger dama mhorr*	Condensed into two subspecies by P. Drüwa 1985 *Gazella dama dama* From eastern and central Sahel *Gazella dama mhorr* From West African mountains and grass steppes

Note: Other scientific names in the literature include *Antilope nanguer* and *Antilope addra* (Bennett 1833), *Antilope dama* var. *occidentalis* and *Antilope dama* var. *orientalis* (Sandevall 1847), *Gazella mhorr reducta* (Heller 1907), and *Gazella dama weidholtzi* (Zimara 1935) (as listed in RZSS and IUCN Antelope Specialist Group 2014).
For citations within the table, see the reference list.

From the early nineteenth century, further forms of dama gazelle–like antelope were described as new species. In 1827, Charles Hamilton Smith named the addra *Antilope ruficollis* based on specimens in Frankfurt (fig. 2.2; see further discussion below). These had been collected by Rüppell in Kordofan, now in Sudan. Not unreasonably, because the dama gazelle had been described from Senegal in West Africa, some 5,000 km (3,105 mi) away, it made sense to recognize this much paler gazelle as a new species. A few years later, in 1833, Edward Bennett described the mhorr gazelle as a distinct species, *Antilope mhorr*, from Morocco. Given the difference in its appearance, this also seemed reasonable. However, even then, Bennett recognized that these three different species might actually form one species, if specimens could be found in intervening areas. Gray (1846) appears to have been the first mammalogist to place the three dama-like gazelles in the genus *Gazella* as *G. dama*, *G. ruficollis*, and *G. mohr*. These three species were recognized until the turn of the twentieth century, when they were united partly by Bryden (1899) and completely by Neumann (1906) into a single species. Several subspecies were also described in the early twentieth century, in part because of the wide morphological variation seen among individuals and also because of confusion about the collecting locality for the original dama gazelle specimen (table 2.1).

In 1984, Mar Cano Perez published her review of the intraspecific taxonomy of the dama gazelle. She recognized three additional subspecies proposed in

A

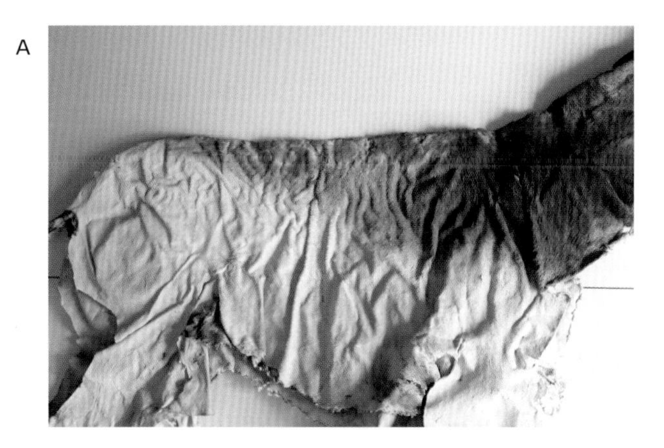

B

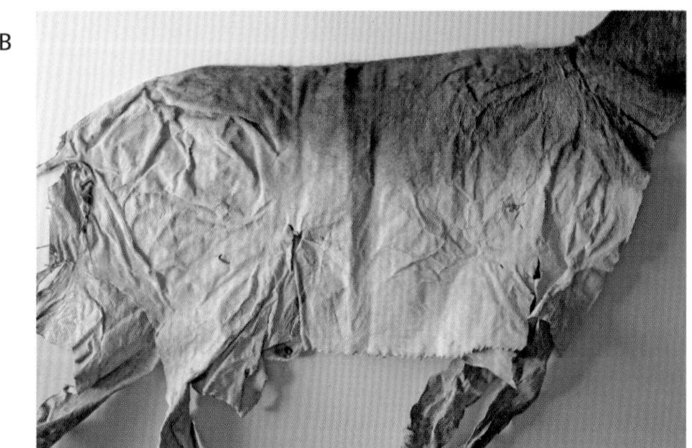

C

Figure 2.2. Type series of skins for *Gazella ruficollis*: *A*, type, male specimen (holotype); *B*, female specimen in the type series (paratype 15920); *C*, young male specimen in the type series (paratype 15884) (photos by A. C. Kitchener, courtesy of the Senckenberg Museum in Frankfurt-am-Main, Germany).

Figure 2.3. Skin illustrating *Gazella dama lozanoi* suggests the dark, extensive coloration of the mhorr gazelle (Morales Agacino 1934, courtesy of Fernando Barroso-Barcenilla, Real Sociedad Española de Historia Natural).

the early twentieth century (*damergouensis*, *lozanoi*, and *permista*; for notes on the original descriptions, see appendix 1) as junior synonyms of the three principal subspecies (table 2.1). For example, the dark and extensive pelage coloration of *permista* and *lozanoi* shows that they should be regarded as junior synonyms of *Nanger dama mhorr*, although *permista* was described erroneously based on a misunderstanding about the type locality of the nominate dama gazelle, *Nanger dama dama*.

Gazella dama lozanoi has a particularly interesting, if brief, history. Cano Perez (1984) referred to it as based on only one incomplete skin, which was acquired from a nomad in Villa Cisneros (now Dakhla), a town on the Bay of Río de Oro, well within the geographical distribution of the mhorr gazelle (Morales Agacino 1934). However, a photograph of the holotype skin shows that it is remarkably complete (fig. 2.3).

Cano Perez (1984) regarded *Gazella dama damergouensis* as a color variant of the nominate *Gazella dama dama* in which the usual white coloration has

Figure 2.4. Dama gazelle in an exotic population on a Texas ranch showing a coat that appears pale and has an unusual yellowish cast, depending on the light (photo by Elizabeth Cary Mungall, courtesy of Kyle Wildlife, Texas, USA).

a yellowish tinge, owing to the way light is reflected by the atypical structure of the hair tips. This color variant is widespread, found principally near the Niger River, but also to a lesser extent elsewhere.

For example, two addra gazelles with similar coloration were reported in a herd of approximately 85 animals on a Texas ranch (Mungall, pers. comm.) (fig. 2.4). All these Texas gazelles are descended from animals (about 20, see chapter 16 on the Chad capture) sent to the United States by Frans van den Brink in 1967 from northeast of Lake Chad.

Defining Subspecies

Cano Perez's (1984) taxonomic review of the dama gazelle, which recognized three subspecies, *mhorr*, *dama*, and *ruficollis*, was based on examination of 50 museum skins and 120 living animals. By plotting collecting localities for each specimen, Cano Perez looked for inconsistencies and gaps in the overall distribution of different pelage patterns (see map 2.1). From this, she concluded that despite the large amount of variability everywhere, coloration became generally paler from the northwest to the south and east, which suggests a cline, but the change in coloration appeared not to be continuous. Each of the three putative subspecies, which are defined morphologically, is associated with a different geographical region. Table 2.2 lists the defining or diagnostic characteristics for each of these three subspecies.

From the geographical distribution of the collecting localities, it was apparent that the dama gazelle is adapted to warm steppe and semiarid regions rather than true desert. The dama gazelle was also found in the massifs of Hoggar and Tibesti in the central Sahara, where conditions are more moderate than in the surrounding desert. Rocky mountains may not be typical dama gazelle habitat, but they may have originally served as a refuge as the climate became drier during the Holocene. Today, they are a vital haven in the face of severe persecution elsewhere (see chapter 20 on the current status of dama gazelles in the wild).

Cano Perez (1984) gave the mhorr gazelle's distribution as being from the Bay of Río de Oro to Senegal. This includes a large region of desert, but its coastal situation means that environmental conditions are not so severe. Cano Perez considered the Senegalese specimens to be identical with those from the northwest, so she classified them as the

Table 2.2. Main distinguishing characteristics for typical western, central, and eastern dama gazelles

1. Mhorr gazelle in northwest, *Nanger dama mhorr*

Reddish-brown and white body color with distinct division basically horizontal.

Color division below center of body.

Neck patch remains small as individual matures.

Nanger rump patch shape clearly defined.

Pigmented area of *Nanger* rump patch continues over haunches and almost to hoofs, expands into "ham" shape on lower haunches, with variable white indentation above the gaskin on the forward edge.

Dark eye stripe reaches nearly to mouth.

2. Dama gazelle in central area, *Nanger dama dama*

Pigmented area continues over haunches but width more restricted so more like a stripe with extension at least to hocks, coloration weak if below hocks.

Division between chestnut and white of body higher than in mhorr and can be somewhat diagonal.

Neck chestnut with color tending to become less intense over the back and sides.

Neck patch usually larger than in mhorr.

No light saddle patch.

Eye stripe shorter and not dark as in mhorr.

3. Addra in east, *Nanger dama ruficollis*

Haunch pigmentation, when present, is like a truncated stripe, not expanded, not reaching hocks, often not connected with main side color.

Color on sides lighter than the chestnut neck, like dark red roan, roan, or light roan, or virtually absent.

Division between roan and white on sides usually distinctly diagonal rather than horizontal.

Neck patch expanded and variable in shape among adults.

Indistinct saddle patch over back of withers common.

No eye stripe or only faint tan hint.

Source: After Cano Perez 1984.

mhorr subspecies. In spite of a 1,200 km (745 mi) gap in available study skins from Senegal eastward to the great bend on the Niger River in Mali, Cano Perez also included all this expanse for subspecies *mhorr*. Not all mammalogists agree that the division between *mhorr* and *dama* is so far east. Cano Perez's decision was probably influenced by the large amount of color variation among the mhorr gazelles bred in Spain at the Almería rescue center, even though they were all descended from only four individuals that were caught in only two limited areas of the former Spanish Sahara (now Western Sahara).

From Niger to western Chad between longitude 7° and 15° east, the specimens examined by Cano Perez (1984) generally have the pelage patterns of *Nanger dama dama*. At longitude 17° east, very close to the eastward limit of the nominate *dama* subspecies, Cano Perez found two or three specimens that matched the dark phase of subspecies *ruficollis* in appearance. Another 800 km (497 mi) to the east are the Sudanese populations, which clearly belong to *ruficollis*. Although Cano Perez found that the dama gazelles from Sudan showed even more pronounced color variation than the *mhorr*, she still regarded *ruficollis* as a well-defined subspecies.

Other mammalogists have arrived at different conclusions. Drüwa (1985) carried out a similar review and concluded that there are only two subspecies:

mhorr in West Africa and *dama* in the central and eastern Sahel, including *ruficollis*. Groves and Grubb (2011) agreed with Cano Perez that there are three subspecies but placed the southern limit of *mhorr* in Western Sahara, so that *dama* occurs from Senegal to west of Lake Chad, with a wide area of intergradation with *ruficollis* in Chad. It is interesting to note that although taxonomists have reviewed similar specimens in Europe's museums, they have arrived at three different conclusions as to the number and geographical distribution of dama gazelle subspecies. This underlines the difficulty of making robust conclusions based on so few specimens and means that other lines of evidence should also be reviewed.

Lines of Evidence

In order to investigate the intraspecific taxonomy of the dama gazelle, we need to have accurate information about the origins of the specimens on which current and past taxonomies have been based. Unfortunately, museum specimens from known localities are scarce and provide only an incomplete picture of former variation, based mainly on trade routes and the places European explorers happened to go. Given the paucity of museum specimens and of records of sightings in early literature, the original extent of the dama gazelle's geographical distribution remains unclear. Therefore, we need to look for additional lines of evidence to try to reconstruct its original range. This may include ecological information, which can tell us with which habitats dama gazelles have been associated; biogeographical information, which can tell us about how climate and habitats have changed over time to influence distribution and isolation of populations; historical information from classical and ancient times; and archaeological evidence, including rock art and skeletal remains found at archaeological sites. A rough approximation might look like the distribution shown by the ovals on map 2.1.

In Roman times, Pliny described the dama as coming from across the sea and said that its horns curved forward, making it a different animal from

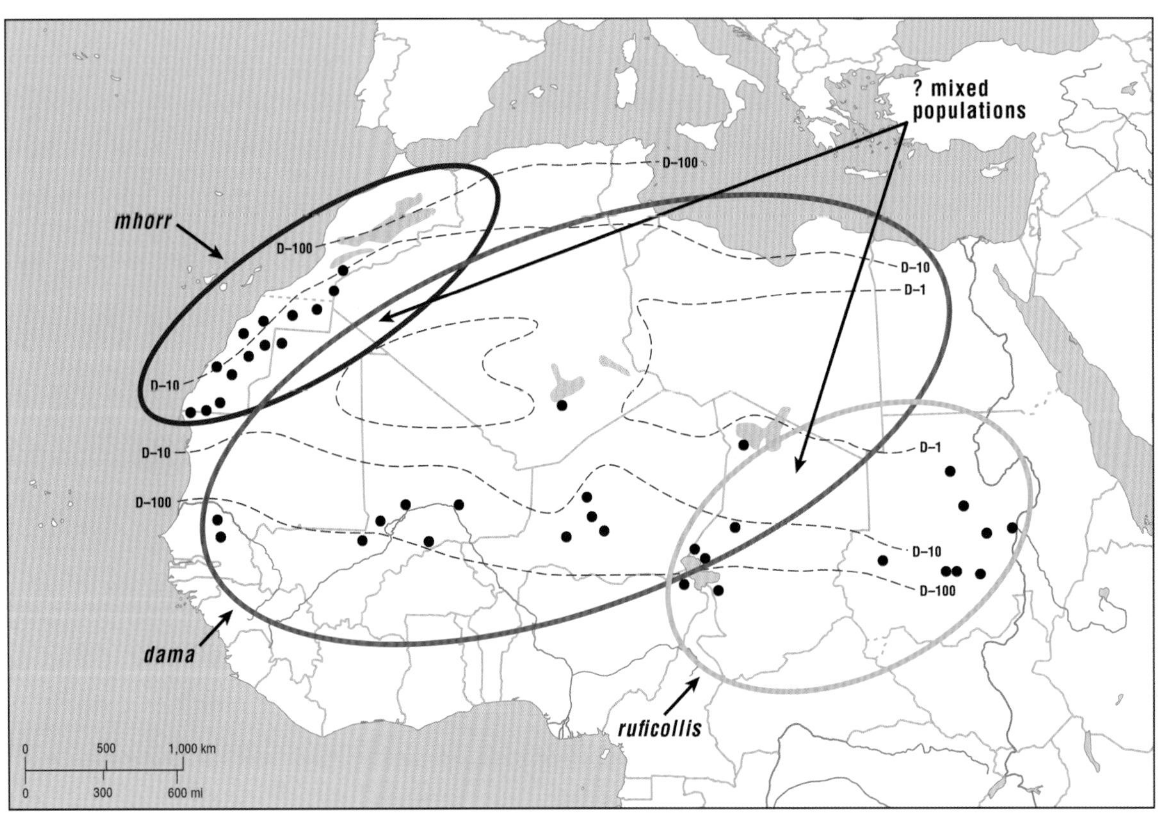

Map 2.1. Subspecies zones overlaid on map by Cano Perez (1984) showing where wild dama gazelles may have formerly lived. Dots represent skins with locality information mapped by Cano Perez. Limits delineated by D-100, D-10, and D-1 indicate progressively drier climate (composite by A. C. Kitchener).

the fallow deer which today has the scientific name *Dama dama* (Bennett 1835). In spite of uncertainty as to what animal Pliny actually meant, in 1764 Buffon identified Pliny's dama with the specimen of the *nangueur* or *nanguer* collected by Michel Adanson from Senegal. In 1766, Peter Pallas gave the dama gazelle its scientific name, *Antilope dama*, based on Buffon's account. (For selections from various early descriptions and a drawing of Buffon's *nanguer*, see appendix 1.) However, there is confusion as to the type locality of the dama gazelle. Buffon stated that Adanson's specimen, which became the type for Pallas's *Antilope dama*, was brought from Senegal (Buffon 1764). Later, Neumann (1906) thought that only the mhorr gazelle, which Bennett had described in 1833, occurred in the west, and consequently he described a new subspecies, *permista*, from Senegal. Certainly, Buffon's *nanguer* appears to be different from Bennett's mhorr gazelle (fig. 2.5). Therefore, Neumann revised the type locality to Lake Chad. As we can see today, captive dama gazelles, descended from stock captured northeast of Lake Chad, often include individuals with haunches as white as those of Buffon's *nanguer* (Mungall 2010). However, Europeans surveyed Lake Chad in only 1823 and so it was still unknown territory for them in the mid-eighteenth century. Could the specimen that Adanson collected and Buffon described have been traded from east of Senegal before Adanson acquired it? Although this is unknown, Adanson (1757) does describe gazelles (although not the species) in his account of his travels in Senegal in the mid-eighteenth century, and he describes hunting these animals himself, so it seems more likely that he collected the specimen personally. His map shows a slightly wider geographical area than today's country of Senegal, and this area may have included animals of more typical nominate dama coloration. Also, Groves and Grubb (2011) have seen museum specimens from Senegal with pelage coloration similar to that of Buffon's *nanguer*, so the type locality remains Senegal, even though we do not know precisely where Adanson's specimen came from.

Figure 2.5. Bennett's mhorr gazelle male as pictured with his 1835 description.

Another important source of information about the past distribution of dama gazelles comes from depictions of animals from ancient times in Egypt and other countries. In particular, cave paintings and other rock art have the advantage of sometimes showing pelage coloration as well as indicating whether the species may have been present locally. The ancient Egyptians depicted several gazelle species in their artwork, although probably not including the dama gazelle (Osborn and Osbornová 1998). The only evidence that Osborn and Osbornová (1998) list for Egypt is dried, unfossilized remains discovered in the Great Sand Sea of the Western Desert (Uerpmann, cited in Osborn and Osbornová 1998) and fossils from the Upper Pleistocene in southwestern Egypt in the Western Desert at Bir Tarfawi and Nabta Playa (Gautier, cited in Osborn and Osbornová 1998). Traveling west from these places where the fossils were found, one would come to Gilf Kebir and Jebel Uweinat, which are mentioned below as dama gazelle rock art sites (figs. 2.6 and 2.7). Whether hunted in or near the Nile Valley or sent as tribute from other regions, dorcas gazelle (*Gazella dorcas*), Persian gazelle (*Gazella subgutturosa*), Soemmerring's gazelle (a close relative of dama gazelle), and possibly slender-horned gazelle (*Gazella leptoceros*) have each been distinguished in ancient Egyptian art. All gazelles were referred to by the same name except for Soemmerring's gazelle, which was often given its own name (Osborn and Osbornová 1998). In her review of gazelles in ancient Egyptian art, Strandberg (2009) shows no representations of dama gazelle, whereas

Figure 2.6. Dama gazelles (with reddish color only on the neck and a narrow strip along the back) in cave painting in southwestern Egypt—hunting scene from the Cave of Beasts, Wadi Sora North, western Gilf Kebir, southwestern Egypt (photo by András Zboray from Zboray 2009).

Figure 2.7. Dama gazelles (reddish color appears to end on croup, and red-white border on sides may be somewhat diagonal) in cave paintings at Jebel Uweinat in the border zone where Egypt, Libya, and Sudan meet. This section from a hunting scene depicts five dama gazelles (photo by András Zboray from Zboray 2009).

Figure 2.8. Gazelle showing well defined *Nanger* rump patch in rock art from the Manga in western Chad (photo by Ursala Steiner from original location in Africa).

she documents that Soemmerring's gazelle was represented often enough to suggest that it occurred in hunting grounds near the Nile Valley. Even so, carvings and paintings of Soemmerring's gazelle are not nearly as numerous as representations of dorcas gazelle, addax (*Addax nasomaculatus*), Nubian ibex (*Capra nubiana*), scimitar-horned oryx (*Oryx dammah*), and bubal hartebeest (*Alcelaphus buselaphus*) (Osborn and Osbornová 1998, Strandberg 2009).

To the west of the Nile Valley, dama gazelles are depicted in cave paintings from deep in the desert in southwestern Egypt at Wadi Sora North, western Gilf Kebir, and Jebel Uweinat, where present-day Egypt meets Libya and Sudan (figs. 2.6 and 2.7) (RZSS and IUCN Antelope Specialist Group 2014). While the gazelles at Wadi Sora resemble addra gazelles, those at Jebel Uweinat look more like nominate dama gazelles. The climate in this region was more moderate during the time of the early Egyptians, but it grew progressively drier. Another 300 km (186 mi) or more to the southwest, similar cave paintings of dama/mhorr gazelles have been discovered at Ennedi, Tibesti, in Chad. Figure 2.8 shows a representation with a clear *Nanger* rump patch from the Manga region in northwestern Chad (RZSS and IUCN Antelope Specialist Group 2014). Rock art at Taghit, Algeria, depicts gazelles that look like addra, while at Tadrart, in Tassili n'Ajjer National Park in southeast Algeria, a mhorr-type gazelle is shown. Other rock art at Iheren (Eherene), central Tassili, shows four typical dama gazelles, and at Rekeiz, Western Sahara, a mhorr gazelle is depicted in a cave painting. Therefore, the images of dama gazelles from rock art show a much more mixed picture of morphological forms, which does not necessarily match modern distributions of today's putative subspecies. Nevertheless, we must be cautious about assuming that these dama gazelles were necessarily depicted accurately.

Archaeological evidence also confirms a wider geographical distribution of dama gazelles in the Sahara, but of course we cannot know what they looked like. Most of the sites are in the eastern half of the Sahara, owing to high levels of archaeological exploration there, and the most easterly sites (e.g. Upper Atbara) probably included the remains of Soemmerring's gazelles, but it is clear that dama gazelles had a more northerly distribution in the early to mid-Holocene in the Sahara. During this period, there was significant climatic amelioration in the Sahara (Larrasoaña et al. 2013), which would have enhanced the potential for gene flow across this region.

Moving on to ecological and biogeographical evidence, we can use the habitats the dama gazelle is associated with as a proxy for its former distribution. However, it is important to note that the current distribution of habitats may be quite different from their extent in the past, owing to far-reaching climatic changes that have occurred over the last 20,000 years since the peak of the last glaciation.

Estes (1991) reviewed the ecological relationships among different gazelle species, whose distributions often overlap. He showed that dama gazelles are less adapted to deserts than other species, for example dorcas gazelles, and hence are associated with the steppe and savanna habitats of northwest Africa and the Sahel. Before ancient civilizations developed along the North African coast and established settlements from Carthage to Alexandria, were dama gazelles found in the northern Sahara too? Dama gazelles enter the Sahara mainly when rains produce a flush of vegetation, but because of their larger size, dama gazelles are more vulnerable to hunting than the smaller slender-horned and dorcas gazelles. This vulnerability may have been exacerbated by the narrowness of the band of North African steppe vegetation as the climate became increasingly dry from the time of the mid-Holocene, compared with the much broader Sahelian savanna belts on the southern fringe of the Sahara (see chapter 1 on the decline of the dama gazelle). Reconstructions of habitats in the Sahara during the Holocene indicate that the dama gazelle would have had a much more extensive geographical distribution than it does today (Larrasoaña et al. 2013), perhaps encompassing most of what is now the Sahara Desert.

If there really are three subspecies of dama gazelle, we would expect to find climatic or geological events that might explain the current distribution of more or less distinct forms caused by population fragmentation. During the Last Glacial Maximum, the peak of the last Ice Age some 20,000 years ago, the deserts were slightly more extensive than they are today, pushing the Sahelian savanna vegetation farther south. Whether this led to populations of dama gazelles being stranded in different noncontiguous patches of suitable habitat on the fringes of a slightly more extensive Sahara is unknown, but it might have led to a separation between what we now call the mhorr gazelle in the north and the southern dama gazelles. This may have provided sufficient separation between populations for morphological differentiation to have occurred, but it is unclear whether there was differential selection across the Sahelian region that led to the apparent pattern of increasingly lighter pelage coloration from west to east. As yet, genetic data do not support this scenario.

Since the last Ice Age and during what we call the Holocene, Lake Megachad, forerunner of today's much smaller Lake Chad, was a potential biogeographical barrier between putative *ruficollis* and *dama* subspecies (maps 2.2*A* and 2.2*B*). Drake and Bristow (2006) estimated that the maximum extent of Lake Megachad was 361,000 ± 13,000 km^2 (158,969 ± 5,200 mi^2) in the early Holocene, 7,500–6,950 BP. A first glance at map 2.2*B* suggests that Lake Megachad could have restricted gene flow between eastern and

A

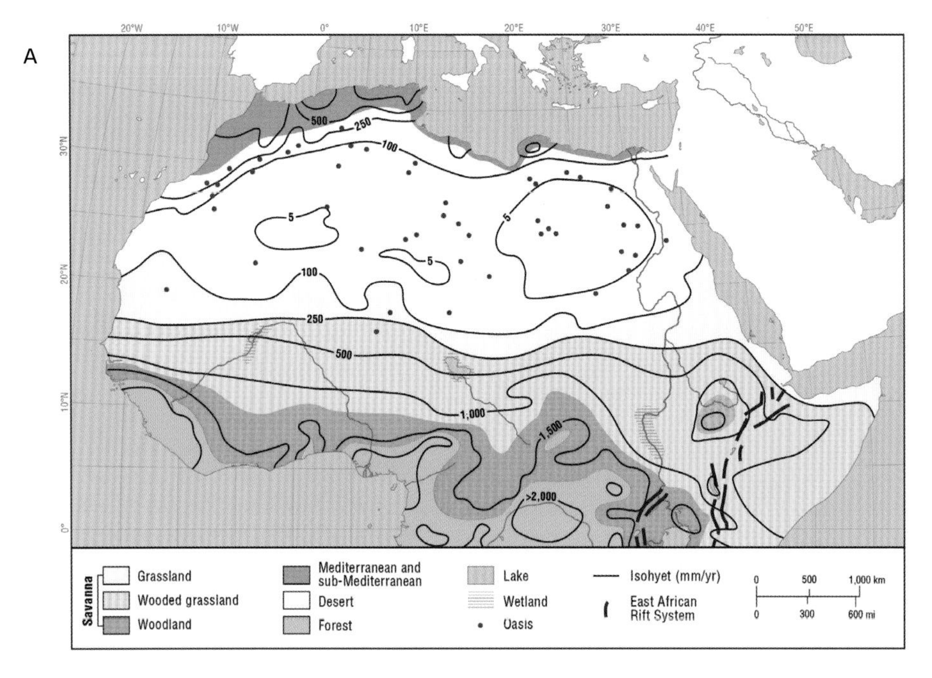

B

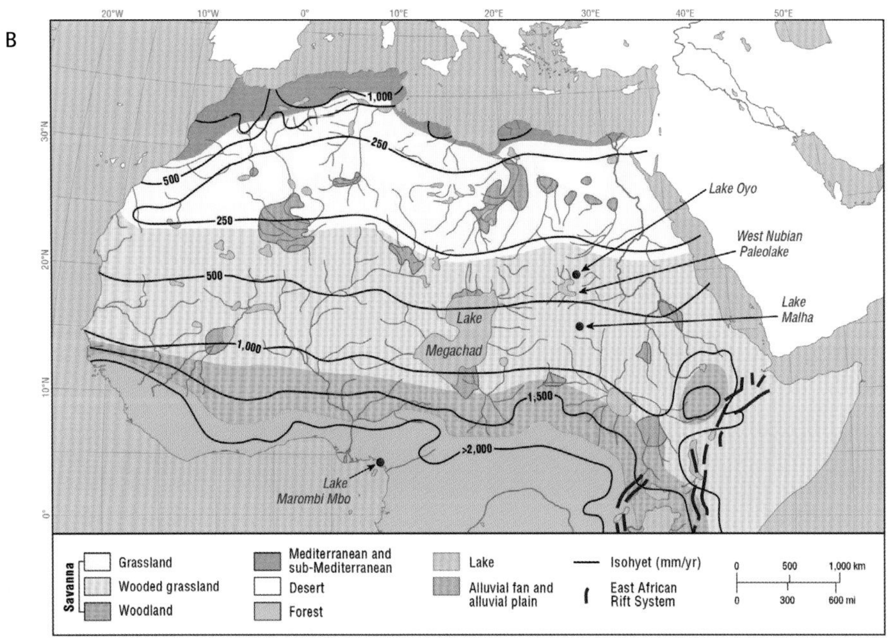

Map 2.2. Vegetation maps for North Africa (by Juan C. Larrasoaña from Larrasoaña et al. 2013): *A*, present-day vegetation; *B*, reconstructed vegetation in the early- to mid-Holocene, 10,000 to 6,000 BP.

central populations, but this assumes that the Sahara was a desert. For Lake Megachad to exist, there must have been much more water available, so the Sahara was mostly not a desert at this time. Reconstructions of the vegetation in North Africa in the early to mid-Holocene, 10,000 to 6,000 BP (Larrasoaña et al. 2013), show that there were probably only a few small patches of arid desert and that most of the Saharan region was benign for dama gazelles, including savannas (map 2.2*B*). This is also confirmed by Neolithic rock art, which shows giraffes, rhinoceroses, and elephants that are associated with these habitats today but that now occur much farther south. The increased monsoons that led to this climatic amelioration lessened during the later Holocene, resulting in an increasingly drier environment. By 4,000 BP, Lake Megachad had shrunk and split into three: Lake Chad, Lake Fitri, and Lake Bodélé. Since at this time the Sahara also had other large lakes, Drake and Bristow (2006) speculate that these water bodies may have resulted in a humid corridor, which would have allowed greater movement of animals across what is now desert until about 4,000 BP. In other words, there was unlikely to have been any barrier to gene flow among dama gazelles during this wetter, milder period of the Holocene.

There are clues about the former mobility of dama gazelles across the Sahara from the high variability in coat color pattern in some regions. For example, pelage variation in captive animals originating from Frans van den Brink's capture at Ouadi Haouach (16°23'22" N, 19°37'54" E; see chapter 16), northeast of Lake Chad, may throw doubt on the whole three-subspecies scenario. Nevertheless, only one adult in 10 years of observations by Elizabeth Cary Mungall of a herd of van den Brink descendants, which grew to more than 85 individuals, ever showed even a faint haunch stripe extension to the hocks, a clear extension being a distinguishing feature for *N. d. dama* (see fig. 2.1*B* and chapter 8). Indeed, pelage patterns of captive animals derived from this imported stock from the same location show that variation in the extent of reddish dorsal coloration can vary even between individuals produced by the same sire and dam, ranging from a virtually white-bodied red-necked gazelle to an animal with almost as much of the reddish color as in a typical *N. d. dama* (Mungall 2010). A similarly high degree of variation was also described for the wider region around Lake Chad. In the 1930s, Malbrant (1952) stressed the lack of uniformity of pelage patterns among dama gazelles. He was skeptical as to the existence of any subspecies, suspecting that pelage patterns represented merely individual variability. He recorded many coat pattern variations around Lake Chad, both in the same herd and in different herds, as well as in different regions. For example, some specimens collected in the region of Rig-Rig, on the east bank of Lake Chad, were as light as Hamilton Smith's red-necked gazelle (see below), while others were similar to the light phase of mhorr gazelle. Specimens taken by Major Powell-Cotton in 1925 (fig. 2.9), collecting northeast of Bou Foumine, which is northeast of Rig-Rig, illustrate some of the variation astride the boundary of longitude 15° east proposed by Cano Perez (1984) as the dividing line between *dama* and *ruficollis* (darker male taken at 14°55'N, 16°13' E, lighter female taken somewhat farther north at 15°5' N, 16°20'E; information courtesy of Inbal Livne, head of collections,

Figure 2.9. Dama gazelle mounts (next to addax mounts) in the Powell-Cotton Museum, United Kingdom, from northeast of Lake Chad. Both of these dama gazelles have the long vertical extension of the haunch stripe as in *N. d. dama*, although the thickness differs. The female (in front) has a white face and no eye stripe (specimen NH.NH.72). The male (behind) has a gray nose patch and tan eye stripe (specimen NH.NH.75). (Photo courtesy of the Powell-Cotton Museum.)

Powell-Cotton Museum, United Kingdom). Both animals have the long vertical haunch stripe extension characteristic of *N. d. dama.*

Going east to Biltine (14.30° N, 20.53° E), within the addra gazelle's presumed distribution, Malbrant saw skins similar to those of the nominate subspecies, and even some that were like the light phase of the mhorr gazelle. I have examined the series of skins collected by Rüppell from Kordofan, including the type, which are still in the collections of the Senckenberg Museum in Frankfurt-am-Main (fig. 2.2). The adult male holotype skin with its limited back color and white haunch is, unsurprisingly, typical of *ruficollis.* The female and juvenile male skins have somewhat more back color and a faint horizontal section of haunch stripe, as on many dama gazelles today called *ruficollis.* This female and young male specimen lack the clear horizontal-plus-vertical stripe to the hocks typical of *N. d. dama* (see figs. 2.1 *B, C*; drawings from Cretzschmar 1826 of these same three animals, reproduced in chapter 4; and discussion on color in chapter 8). These specimens were all from the same locality. Clearly, the simple clinal pattern of variation that has been assumed for the dama gazelle was much more complex, suggesting instead that variation in pelage coloration followed a roughly northwest-to-southeast transition (map 2.1) rather than a west-to-east one, if it can be supported at all.

Therefore, it is probable that the pelage patterns of different putative subspecies occurred together over a wide area of the Sahara over the last 10,000 years. Subspecies should be recognized only where different forms are geographically, as well as mostly genetically and morphologically, distinct. In other words, we would not expect putative subspecies to be sympatric, which was clearly not the case even in the recent past.

A recent crossbreeding experiment at Al Ain Zoo in the United Arab Emirates produced 10 surviving animals from matings between mhorr and addra gazelles (Anonymous 2016). These so-called hybrids bear a strong resemblance to nominate dama gazelles, *N. d. dama,* or some of the darker addra gazelles. Perhaps these breeding experiments offer a strong clue as to how morphological variation has arisen within *Nanger dama.* If during the last or a previous glaciation, dama gazelles became isolated as two populations, which, owing to drift or unknown selection, diverged morphologically into the extremes of a mhorr-type and addra-type gazelle, they could have then expanded their range during the early to mid-Holocene when climatic conditions were optimal across the Saharan region, allowing the high levels of gene flow that gave rise to the lack of genetic structure in populations today. As a result of this mixing, involving a complex series of crosses and backcrosses, we see a mixture of morphological forms (as observed by Malbrant 1952 and others) within a general cline from dark to light gazelles, running from west to east. Therefore, the so-called subspecies are merely an artifact of limited sampling in increasingly fragmented populations driven southward by deteriorating climatic conditions during the later Holocene. Further experimental breeding, including the production of F_2 crosses and backcrosses of F_1 animals to parental forms, would add further insight into this possible scenario.

Conclusion

Further research involving multiple lines of evidence is clearly required in order to determine the meaning of the geographical variation in dama gazelles. Currently, the evidence is not strong for recognizing any subspecies within the dama gazelle, but there are intriguing differences in morphology that apparently contradict genetic and biogeographical data that suggest high levels of gene flow over the last 20,000 years. So, what is a dama gazelle? The picture is much more complex than previously realized, but an answer is urgently needed, because time is fast running out for the dama gazelle both in the wild and in captivity. There are so few dama gazelles surviving today that we may have to be pragmatic about their future conservation in order to minimize inbreeding and prevent the fixation of deleterious alleles in both wild and captive populations, regardless of what the animals look like.

Chapter 3

Conservation Implications from Genetic Testing

Helen Senn

The only way to gain a clearer understanding of the level of differentiation between the reputed subspecies of dama gazelle is to conduct an extensive genetic and morphological survey of wild populations, both contemporary and historical. Historical studies were based on relatively limited numbers of specimens and were not able to use DNA analysis, which is today considered to be a mainstay of taxonomy (for a review of historical taxonomy, see chapter 2). However, the technical challenges of such an ideal study would be considerable, not least because many populations of dama gazelle are simply no longer in existence, but also because of the difficulty of gathering a sufficient number of high-quality genetic samples from them.

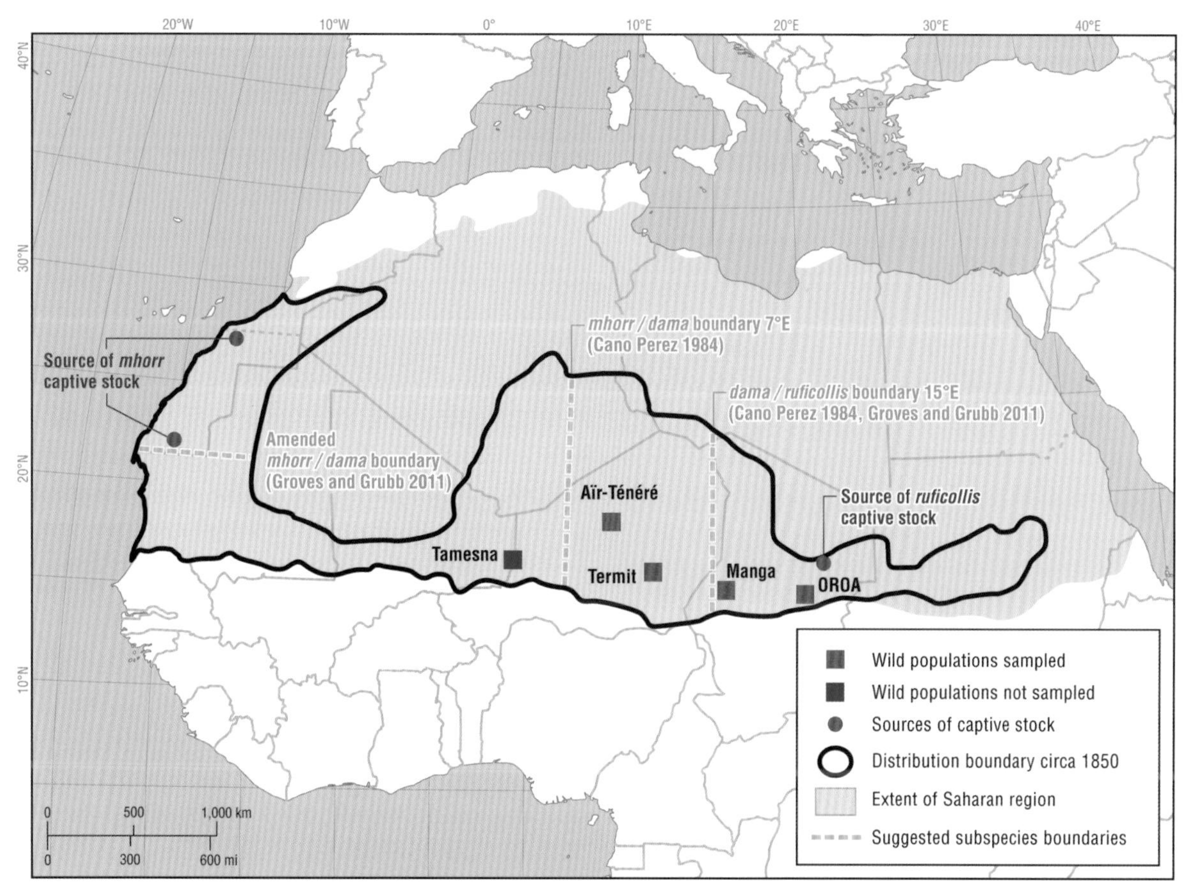

Map 3.1. Former and current distribution of the dama gazelle (adapted from Durant et al. 2014 and Jebali 2008), including sites of the known remaining wild populations and subspecies boundaries proposed by Cano Perez (1984) and by Groves and Grubb (2011). Origins of captive populations and extent of sampling for genetic diversity indicated. See text for details. (Map modified from RZSS and IUCN Antelope Specialist Group 2014.)

Table 3.1. Dama gazelle populations and samples

Population	Site code[1]	Details	Putative subspecies	Total number of samples collected
WILD				
Chad (Ouadi Rimé–Ouadi Achim)	OROA_R	Wild population in Ouadi Rimé–Ouadi Achim Game Reserve in central Chad (~ 14.9027 N, 19.8318 E)	*N. d. ruficollis*	18
Chad (Ati)	ATI_R	Ati locality (~ 13.11 N, 18.20 E)	*N. d. ruficollis*	1
Chad (Manga)	MANGA_R	Wild population in Manga region of western Chad (~ 15.33087 N, 15.1277 E)	*N. d. ruficollis*	36
Niger (Termit)	TERMIT_D	Wild population(s) in the central (~ 16.1047 N, 11.4171 E) and northern (~ 16.3706 N, 11.4581 E) massif of the Termit Mountains	*N. d. dama*	22
Niger (Aïr)	AIR_D	Wild population in Aïr Mountains (~18.642 N, 9.80924 E)	*N. d. dama*	8
ZOO/CAPTIVE				
Al Ain Zoo *'mhorr'*	AIN_M	Origin unrecorded, highly likely to be descended from animals in the EEP[2] (originally from EEZA[3]).	*N. d. mhorr*	42
Al Ain Zoo *'ruficollis'*	AIN_R	Origin unrecorded, likely to stem from the North American Regional Studbook for addra (*ruficollis*) gazelle, as it records the transfer of two females and a male to Al Ain Zoo in 1982 (Petric 2012).	*N. d. ruficollis*	20
Dama gazelle *'mhorr'* EEP	EEP_M	Animals sampled from City of Belfast Zoo, EEZA, and Montpellier, all ultimately originating from EEZA	*N. d. mhorr*	12
Marwell Zoo *'ruficollis'*	MAR_R	Origin is the North American Regional Studbook for addra (*ruficollis*) gazelle (Petric 2012).	*N. d. ruficollis*	5
Katané, Ferlo North Game Reserve, Senegal	SEN_M	Ultimately originating from EEZA via Réserve Spéciale de Faune de Guembeul, Senegal.	*N. d. mhorr*	3
Safia Reserve, Morocco	SAF_M	Ultimately originating from EEZA via R′Mila Royal Reserve, Morocco.	*N. d. mhorr*	6
USA captive	USA_R	Samples from White Oak, Fossil Rim, Kyle Wildlife, Dallas Zoo, Jackson Zoo, Metro Richmond Zoo, and previously published data from San Diego Zoo (Hassanin et al. 2012),[4] all reportedly descended from animals from OROA.[5]	*N. d. ruficollis*	53 (1)[4]

[1]The last letter of each site code designates the putative subspecies of dama gazelle represented by that population: R = *ruficollis*, D = *dama*, M = *mhorr*.
[2]EEP = European Endangered Species Programme.
[3]EEZA = European Association of Zoos and Aquaria.
[4]Data reported by Hassanin et al. (2012) are from a single specimen.
[5]OROA = Ouadi Rimé–Ouadi Achim Game Reserve.

The first, and to date only, genetic study on wild dama gazelle populations examined fecal samples collected during field surveys conducted by the Sahara Conservation Fund between 2009 and 2016 (Senn et al. 2014, 2016). As part of this study, 85 samples were analyzed from across three of four known populations and one additional locality each in Chad and in Niger (table 3.1), meaning that all but one of the extant known wild dama gazelle populations were sampled (map 3.1). Samples were also collected from captive populations within Europe, the United States, the Arabian Peninsula, and Africa. All these animals are known to have descended either from the original founders captured in Western Sahara (*mhorr*) or from the animals captured in eastern Chad (*ruficollis*) (table 3.1 and map 3.1; see also the chapters in sections 3 and 4). Samples were examined using two genes from the mitochondrial genome commonly used in taxonomic studies (cytochrome B and control region/d-loop). Mitochondrial DNA (mtDNA) is present in high copy numbers in each cell, making it particularly suitable for work on degraded DNA such as that from fecal samples. It is passed only from mother to offspring, thus giving an evolutionary view of the species matriline (maternal line). The resulting genetic data were used to examine two separate issues:

1. Genetic diversity within dama gazelle populations globally
2. Dama gazelle taxonomy

Genetic Diversity of Dama Gazelles

The study discovered a total of 37 mtDNA matrilines (haplotypes, groups of genes inherited together from one parent) within the data set of 227 samples. The vast majority (29) of these were found to be within the wild samples. The captive populations, descended from *mhorr* and *ruficollis* founders, possessed 2 and 5 mtDNA matrilines, respectively. This illustrates clearly the genetic consequences of the bottleneck that these populations went through following capture (see chapters 16 and 17 for details) and also likely the genetic effect of subsequent inbreeding in captivity. Even though more material was available from the United States than from any of the other types of locations or wild sites sampled, the US submissions still represent only a small percentage of the captive *ruficollis* there. The number of *ruficollis* tallied for 18 North American institutions registered with the International Species Information System (now renamed Species360) was 220 (RZSS and IUCN Antelope Specialist Group 2014), and more than 1,000 *ruficollis* are held by private owners in the United States, mainly on Texas ranches (see chapter 15 for more about Texas ranches). It may be that further genetic diversity from the original founder population exists among them.

It is no surprise that even so far, the present US material shows greater gene and nucleotide diversity than the captive *mhorr* sample from Europe, since all *mhorr* in captivity originate from only four founders.

Table 3.2. Some genetic diversity statistics for the dama gazelle

Site code*	Number of samples	Number of matrilines/ mtDNA haplotypes	Number of polymorphic sites	Genetic diversity	Nucleotide diversity
USA_R	53	5	47	0.623 ± 0.053	0.01183 ± 0.00592
EEP_M	12	2	17	0.471 ± 0.082	0.00558 ± 0.00303
MANGA_R	32	14	66	0.921 ± 0.027	0.01464 ± 0.00736
OROA_R	18	8	60	0.856 ± 0.055	0.01450 ± 0.00747
AIR_D	8	4	50	0.750 ± 0.139	0.01374 ± 0.00767
TERMIT_D	22	3	27	0.628 ± 0.060	0.00663 ± 0.00351

Note: Higher measures are considered to be positive, whereas lower measures indicate restricted genetic diversity.

*Last letter of each site code designates the putative subspecies of dama gazelle represented by that population: R = *ruficollis*, D = *dama*, M = *mhorr*.

By contrast, 20 *ruficollis*, mainly if not all captured from different herds, were divided right away between two different institutions after being shipped to the United States (chapter 16 also mentions where in Europe the other original van den Brink dama gazelles were lodged).

The data set also illustrates subsequent bottleneck events for the *mhorr* and *ruficollis* captive populations through transfer of individuals across the world. As figure 3.1 illustrates, only a single haplotype was discovered in the population of *mhorr* at Al Ain Zoo (AIN_M), at Safia in Morocco (SAF_M), and at Katané in Senegal (SEN_M)—although note that the sampling of the latter two sites was limited—and only a single haplotype in *ruficollis* was found at Al Ain Zoo (AIN_R) and at Marwell Zoo (MAR_R), although only a limited number of samples were available from Marwell. This suggests that management practices are causing localized reductions in genetic diversity and that the dama gazelle community should give serious consideration to ways to maximize genetic diversity in future reintroductions and transfers. At least Texas ranch owners raising *ruficollis* maintain a policy of circulating animals among each other in order to reduce inbreeding. This practice used to include US zoos until philosophical changes within the zoo community virtually separated these two segments of the US dama gazelle population (see chapter 14). Many people concerned about the ultimate viability of the species are working to restore the former ranch-zoo interchange.

The Conservation Centers for Species Survival (C2S2) Source Population Alliance (see chapter 11) is a positive step toward improving management practices for the species in captivity and toward creating a coordinated approach that includes genetic resources present within both private collections and zoos in the United States. Similar initiatives might also be possible on the Arabian Peninsula, where there are also a considerable number of animals across diverse institutions (see chapter 13). Logistical constraints and legislation are likely to continue to be serious barriers to gene exchange among dama gazelle populations from a global perspective, but it would be advantageous to explore avenues for global gamete transfer.

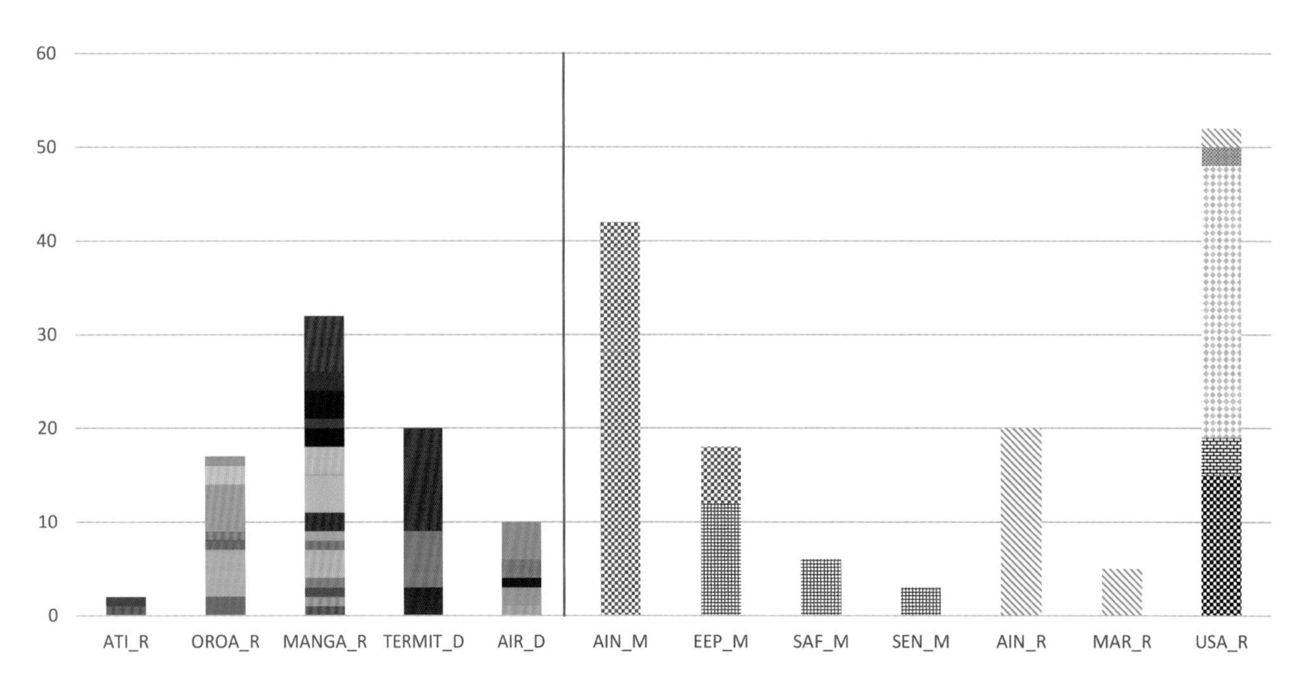

Figure 3.1. Bar chart of control region haplotypes found at different sampling sites in this study. Sites are enumerated by collecting locality (see table 3.1) and a letter indicating putative subspecies; R = *Nanger dama ruficollis*, D = *N. d. dama*, and M = *N. d. mhorr*. Wild populations are shown to the left of the red line, captive populations to the right. Particularly striking, but not unexpected, is that haplotype diversity is higher in most of the samples from wild populations than it is in samples from captive and captive-derived *mhorr* populations and most of the captive *ruficollis* populations. No haplotypes are shared between wild populations, or between wild and captive populations.

The picture for dama gazelles in the wild is quite different, with large numbers of mitochondrial DNA matrilines and strong partitioning of them between different localities (fig. 3.1). Sadly, we will never know, of course, how this genetic diversity would have compared to that of the original wild population. Although genetic diversity in captivity is generally lower than in the wild, the level of genetic diversity uncovered in the US population is so far comparable to that found in the wild population in Termit. While this is a marginally positive sign for the US captive population, it signals another depressing sign for the wild population in Termit. Despite the "relatively high" numbers of dama gazelles in Termit compared to the numbers in other wild populations (RZSS and IUCN Antelope Specialist Group 2014), it could be that the Termit population has undergone a bottleneck—in other words, that it issued from a small group of founders. Today's population of roughly 50 individuals is located in marginal habitat for the species (more or less at the edge of the potential distribution in terms of rainfall) and has managed to survive, while others that were distributed in the original habitat (fixed dunes with vegetation) have been killed because of hunting or have died because of drought.

Populations of threatened species should be managed to maximize genetic diversity and to minimize inbreeding for two reasons: to reduce the likelihood of inbreeding depression, and to retain the greatest adaptive potential of the population. Inbreeding has been shown on numerous occasions to have a detrimental effect on fitness in naturally outbreeding species (Darwin 1876, Crnokrak and Roff 1999, Spottiswoode and Møller 2004, Frankham 2010). This includes studies of captive populations (Ralls and Ballou 1983) and populations released into the wild (Frankham 1995, Kephart 2004, Vilas et al. 2006). Inbreeding depression is not the only reason to be concerned about inbreeding in dama gazelles. Concurrent loss of genetic diversity may result in loss of adaptive potential, limiting the ability of the population to evolve (Frankham et al. 2002, Swindell and Bouzat 2005, Latta 2008), to be resilient to environmental change or to disease (Sgrò et al. 2011, Weeks et al. 2011), and to readapt to wild environments from captivity (Araki et al. 2007, Frankham 2008, Christie et al. 2012).

Although genetic diversity in the wild appears to be considerably higher than in captivity, diversity is undoubtedly being lost in wild populations also. Populations are small and isolated, and in the long term they may become vulnerable to inbreeding if the situation remains unchanged. A critical task for dama gazelle conservationists is to secure both in situ and ex situ genetic diversity and population connectivity. However, hunting, habitat loss, and competition for grazing as a result of pastoral development, as well as inherent vulnerability to demographic fluctuations caused by small population size, are more immediate threats to the survival of the species than loss of genetic diversity per se.

Taxonomy of the Dama Gazelle

A crucial task for the genetics study was to make an assessment of the relatedness of dama gazelle populations. To do this, the pattern of relatedness of the mitochondrial DNA matrilines was examined with respect to geography.

Genetic relatedness in the mitochondrial DNA data was in fact not associated with geographical location or with traditional subspecies designations (fig. 3.2). Closely related haplotypes (groups of genes inherited together from one parent) are found across wide geographic areas, and conversely, haplotypes from specific populations or subspecies do not group closely together. For example, in figure 3.2, the *mhorr* haplotypes can be found closely grouped with those from *ruficollis* and *dama* populations. This strongly suggests that these three traditional subspecies designations are not valid (for more detailed genetic arguments, see Senn et al. 2014, 2016).

Although some Robertsonian translocations (a kind of variation in the number and arrangement of chromosomes) have been found between captive *mhorr* and *ruficollis*, it seems unlikely that they would cause reproductive incompatibility since these sorts of translocations are common in ungulates (Effron et al. 1976, Arroyo Nombela et al. 1990, Vassart et al. 1993). Given the low founder base of captive populations, care must also always be taken to avoid over-interpreting apparent differences among subspecies based on study of captive individuals alone. A number of other lines of evidence based on past and current phenotypic variation (variation based on what the animals look like) and likely history of

Figure 3.2. Relatedness of matrilines/haplotypes. Each haplotype is a circle color-coded according to population of origin, and single base-pair step-wise mutations between haplotypes are denoted by tick marks. Haplotypes with fewer genetic differences are more closely related (i.e. II and IH are more closely related than IH and IAK). The fundamental structure of the haplotype network is unresolved, with a number of possible arrangements being drawn (represented by the loops in the network). However, it is clear from the arrangements of the terminal clusters that the network does not follow any geographic structure. For example, the "*mhorr* haplotypes" IM and IF group closely with haplotypes from a number of "*ruficollis*" and "*dama*" populations.

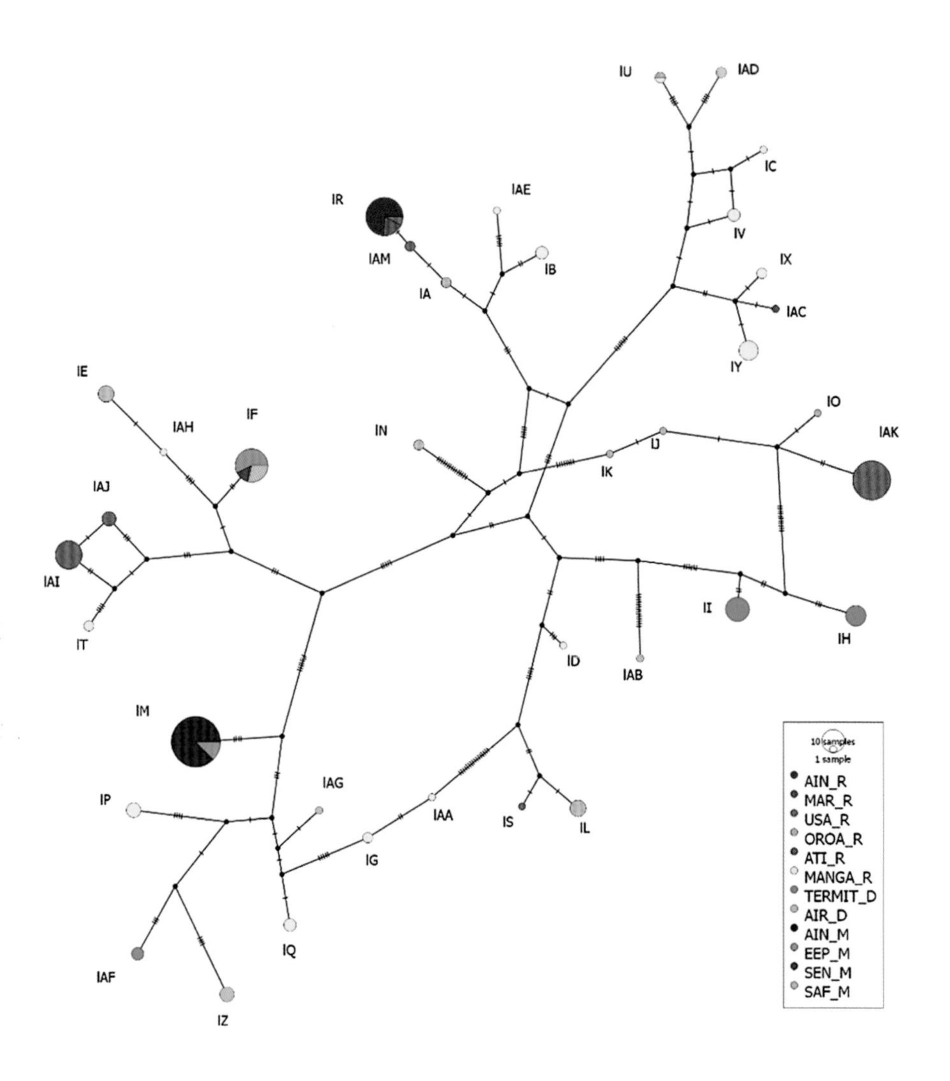

the species (see chapter 2) also support the claim that subspecies designations are not valid for dama gazelles.

The problem with the historical preference for splitting species into subspecies is that the burden of proof rests with the contemporary researcher. The null hypothesis becomes "there are *n* subspecies," even if the naming of subspecies was not done on a rigorous basis (by today's standards). A large number of putative dama gazelle subspecies were recognized historically, based on rather limited evidence, which the paper by Cano Perez (1984) rationalized to only three based on the best available phenotypic and geographical data at the time. It is arguable that in retrospect, perhaps even this was not far enough owing to the apparent clinal and nondiscrete nature of the variation. Today, although it can always be argued that higher-resolution genetic data (e.g. thousands of nuclear genes) would be required for final confirmation, the evidence at the level of mitochondrial genes is suggestive, and it is important to note that the genetic data would not support a split into three subspecies from an a priori assumption of one.

Conservation conclusions based on the genetic data include the following:

1. There is no a priori reason to divide the cline into three discrete units, and the lack of coincident mtDNA genetic structure supports this view.
2. The conservation of the dama gazelle will be greatly advanced if it is considered a single species without subspecies divisions, even though it exhibits phenotypic variation. Under the "three subspecies view," artificial, impermeable boundaries are erected, and only dama gazelles that originate from within the same assumed subspecies geographic area should be bred together and used for reintro-

duction and population augmentation. Under the "monotypic species view" there is a continuum of suitability of donors to a population, where, all else being equal, the geographically most proximate population is the most suitable, but there is no a priori barrier to exchange between any populations provided the risks of exchange have been evaluated properly.

3. Inbreeding depression and loss of genetic diversity must be taken seriously during ongoing management efforts. Evidence for inbreeding depression in a captive population has already been found for the mhorr gazelle, for which there were only four original founders (Ruiz-López et al. 2012). Captive populations of dama gazelles should be managed to maximize genetic diversity and minimize inbreeding. To achieve this aim, continual and improved coordination and monitoring of captive breeding efforts across the world are vital.

4. Unless there is evidence to the contrary, phenotype should be assumed to be under some degree of selection in the wild, as this is the most conservative scenario for conservation. However, no attempt should be made to breed or select for a "true phenotype" in captivity, as this will result in further loss of genetic diversity and possibly unintended selection for traits linked to phenotype.

5. The risks associated with breeding animals from geographically discrete distant populations with each other should be taken seriously, and the mixing of putative subspecies should be evaluated properly as a conservation tool for the dama gazelle. Highly controlled and scientifically monitored experimental crosses of captive *mhorr* and *ruficollis* individuals are being conducted to evaluate whether outbreeding will be problematic, before the implementation of any full-scale genetic mixing of captive populations, before any augmentation of wild with captive populations originally sourced from distant geographic regions, and before any exchange between distant wild populations (Latta 2008). These studies are currently beginning at Al Ain Zoo (see chapter 2).

Chapter 4

Artwork through the Ages

Elizabeth Cary Mungall and Abdelkader Jebali

The following set of pictures of dama gazelles shows a succession of styles characteristic of different centuries. Other than bones, rock art gives the first record of the kinds of animals that were familiar to people in various regions. By the 1700s and 1800s, increasingly refined pictures shared with a wider audience what travelers were discovering and what scientists were discussing. In today's world, photography has often taken over as the medium of choice. This results in paintings concentrating even more than formerly on their purpose as evocative representations and special-purpose creations.

Figure 4.1. Dama gazelle in a noose or on a leash. Rock art from Jebel Uweinat (where Egypt, Libya, and Sudan come together), North Africa (photo by András Zboray from Zboray 2009).

Figure 4.2. Young dama gazelle male collected by Michel Adanson from the vicinity of Senegal (*nanguer* drawing used by Buffon 1764).

These are artist re-creations of the specimens that Rüppell collected in Kordofan (now in Sudan). In 1827, Charles Hamilton Smith described *Antilope ruficollis* on the basis of the skins. For photographs of these skins, see chapter 2.

Figure 4.3. *A*, Dama gazelle male; *B*, female and adolescent (Cretzschmar 1826). For actual skins preserved from these individuals, see figure 2.2.

Figure 4.4. Eastern dama gazelle, also known as addra or red-necked gazelle, of the kind named *Antilope ruficollis* by Charles Hamilton Smith. Later the name was changed to *Gazella ruficollis*; now it is *Nanger ruficollis* (Sclater and Thomas 1897–1898).

Western dama gazelle, known as mhorr gazelle, was originally named *Antilope mhorr* by Edward T. Bennett; the name was later changed to *Gazella mhorr* and is now *Nanger mhorr*.

A

B

Figure 4.5. *A*, Mhorr gazelle male sent live to London from the extreme southern coast district of Morocco, opposite the Canary Islands, and later exhibited as a mount in the British Museum (Sclater et al. 1897–1898); *B*, mhorr gazelle female drawn from life and figured by Geoffroy St.-Hilaire and F. Cuvier in their book *Histoire Naturelle des Mammifères* (picture used by Bennett 1833).

Figure 4.6. Dama gazelle and fawn in painting by A. J. McCoy, produced by the World Wildlife Fund and sold to benefit conservation needed in order for the species to survive in the wild (card copyright © WWF-US).

Figure 4.7. Mhorr gazelle and fawn in painting by C. Drummond, produced by the World Wildlife Fund and sold along with commemorative coin and Senegalese stamp honoring the dama gazelle (philatelic-numismatic cover copyright © WWF-US).

Section 2 Overview of a Species

Section 2 explores the behavior of dama gazelles (chapters 5 and 6), their growth and development (chapters 7 and 8), reproduction (chapters 5, 6, and 7), and food habits (chapter 9). Elements of all these topics come together in the final chapter of section 2, on hand raising the dama gazelle fawn (chapter 10).

The majority of the behavior information is new. With native populations dwindling and with dama gazelles getting pushed into increasingly remote regions of the Sahel and edges of the Sahara, study sites and the resources to sustain observation work have become more and more difficult to secure—not least because of problems caused by the civil unrest that has been prevalent in some of these regions. The behavioral data coming out of the wild are tallies of such things as group numbers seen on wildlife surveys as well as movements with the seasons or the rains to find shelter or food. Both in African breeding centers and on Texas ranches, much more can be documented for dama gazelles in rangeland pastures.

Lodging of newly imported dama gazelles in zoos for their permanent quarantine—only their offspring being allowed to go to ranches and other nonquarantine sites—allowed the editor of this volume to study innate behavior such as that involved in courtship and mating and also allowed observations with both original imports and first- and second-generation animals present (Mungall 1980). But zoo enclosures are small and their environments are limited. Not until captive dama gazelles increased in naturalistic rangeland pastures such as those in African reintroduction centers and on Texas ranches did realistic settings for long-term behavior research present themselves.

For the last 12 years, the editor has been following the behavior and development of ranch animals in a series of Texas pastures. The study started with seven known individuals (a herd buck and six adult does) and their first set of births. All of these births were documented and all of these fawns were tracked individually. This group, eventually given 21.6 ha (53.4 ac), expanded to more than 85 animals before a few maturing males were removed to live elsewhere. This, plus shorter investigations in larger pastures (the largest being 8,996 ha [22,220 ac]), GPS-radio collar studies of 12 to 18 months where animals were too shy and too spread out or where extensive stands of heavy brush prevented reliable observation, and lots of invaluable information added by owners, managers, and others familiar with the Texas animals, has provided a wealth of opportunities to gather developmental as well as behavioral data on eastern dama gazelles living more like free-ranging animals.

The editor's observations of some of the few western dama gazelles kept in the United States have been much more limited. During visits to what was then the only Texas ranch with western dama gazelles, the animals in two breeding pastures were observed, photographed, and discussed with the owner. This allowed important comparisons of physical changes with age—including coat color development—as well as of the dynamics of power transfer between herd males.

A detailed understanding of growth stages and reproduction (chapters 7 and 8) is valuable for managers of captive populations, as it allows them to see whether their animals are progressing normally and whether intervention is needed. It is also valuable for conservationists who need to consider what sex, age, and social relationships would be best to keep a wild population going or to use for reintroductions.

As part of the development documentation, a chapter on horns and color is included (chapter 8).

This is not solely because these animals are so intriguing and pretty. This is just as much to show normal ranges of variation for the eastern dama gazelle—the first dama gazelle subspecies to appear in the United States, the one that has become established in impressive numbers, and the one with the greatest color variation. This is also to analyze the way the distinctive coat pattern of the dama gazelle is constituted. Because coat coloring has been the major aspect used to classify dama gazelles, an understanding of its details is important. Besides taxonomists, other people, such as US wildlife inspectors charged with determining, on the basis of what they can see, what kind of animal is in front of them, need details. For example, the vertical part of the haunch stripe in dama gazelles develops as a separate element from the horizontal part. Together, this pair of parts is characteristic of the central dama gazelle. Just the horizontal part, or none at all, indicates the eastern dama gazelle. But the final condition shows only as the animal changes its coat to adult colors.

Food habits (chapter 9) is the one subject discussed in section 2 with major input from native habitat. The authors of this chapter review food habit reports from other investigators and combine these with their own observations. The chapter documents both the foods taken and the feeding strategies of western dama gazelles. Much of this work was carried out in a West African enclosure of 440 ha (1,087 ac) of native rangeland, later expanded to 1,240 ha (3,064 ac). These findings are then compared with information for the eastern dama gazelle on Texas rangeland, as observed during Texas behavior studies.

Behavior, basic biology, and ecology all contribute to section 2. Information on space-use issues such as territoriality and on preferred forage such as acacias is valuable for any assessment of a living site, either captive or native. Anyone involved with considering reintroduction options will want to pay special attention to these concerns.

Chapter 5

Mothers, Fawns, and General Behavior

Elizabeth Cary Mungall

This chapter covers reproductive behavior in many of its forms, including maternal care. It also outlines daily activity and how mothers and growing youngsters synchronize their changing stages with the general activity of the group. Territoriality is covered in chapter 6.

Reproductive Behavior

One of the most important behavior categories involves reproduction. It covers a multitude of aspects. Primary among these are courtship and mating, the birth process, and maternal care.

Courtship and mating

For successful mating, the female must be moving straight ahead of the male, but not too fast. Courtship works to achieve this. It synchronizes the readiness of both parties to mate. Figure 5.1 shows the type of courtship sequence seen most often because it is performed over and over again until the partners are ready to progress.

Figure 5.1. Most common courtship sequence: *A*, male is in erect posture; *B*, male lowers nose to female's back or nudges her forward; *C*, male follows female; *D*, male resumes erect posture when female stops (photos by Elizabeth Cary Mungall, courtesy of Kyle Wildlife, Texas, USA).

By positioning himself behind the female and urging her forward, the male tests her receptivity. Since all females are subordinate to all adult males, the female's inclination when an adult male comes up behind her is to move away. A young female soon discovers that she can distract the male's attention, at least briefly, by pausing and urinating. The male puts his nose near the female's rear and then may lower it near the ground. Next comes *Flehmen*, in which the male holds his head up to the front and then often to the side and opens his mouth and lifts his upper lip, letting chemical stimuli reach his vomeronasal organ (Jacobson's organ) and test the estrous state of the female (fig. 5.2). After performing *Flehmen*, the male either resumes his courtship approach or abandons what is likely to be a lost cause.

An adolescent male harassed by repeated approaches from a dominant male is apt to discover that if he also lifts his tail as he lowers his head in submission and heads away—reminiscent of a female involved in urination for her suitor—then this may deflect the adult male's attention (fig. 5.3).

Subsequent to this *Flehmen* test, the suitor ambles up behind the female, drawing his neck toward the vertical and his head horizontal, as he did in his initial approach. When the female stops, he accentuates his erect posture and may seem to nudge her forward with his chest. Soon, he is likely to perform

A

B

Figure 5.2. Female's delaying tactic: *A*, male lowers his nose to female's rear as she lifts her tail and urinates; *B*, male performs *Flehmen* as female is free to graze or move away (photos by Elizabeth Cary Mungall, courtesy of Kyle Wildlife, Texas, USA).

Figure 5.3. Two adolescents raise their tails like a urinating female as they lower their heads and necks in submission and move away from the approaching adult male (photo by Elizabeth Cary Mungall, courtesy of Kyle Wildlife, Texas, USA).

Laufschlag, either the classic stiff-legged lifting of a foreleg toward her hind quarters (fig. 5.4), or sometimes a bent-legged, lazy-looking version. The female moves on. In another variation, the male moves up beside his intended as they move along together.

Occasionally, the male approaches with head and neck outstretched, ears dropped out to the sides, and emits a soft series of grunts (fig. 5.5). Sometimes, a bachelor will also come up to another bachelor with this outstretched and ears-out posture.

Figure 5.4. Courting male performs *Laufschlag*, raising a stiffened foreleg toward the female (photo by Elizabeth Cary Mungall, courtesy of Kyle Wildlife, Texas, USA).

Figure 5.5. Male approaches his target—female or male—in outstretched posture with ears turned to the sides while emitting a series of soft grunts (photo by Elizabeth Cary Mungall, courtesy of Kyle Wildlife, Texas, USA).

Eventually, the female may run from the male's attentions. A spirited chase follows. The female dashes into any nearby group, but the male stays with her (fig. 5.6). Finally, she slows, and the initial courtship patterns begin again. Thus, early courtship stages occur much more frequently than later stages. As courtship becomes serious, the female may stay closer to her partner as she moves. Soon, mounting starts to appear. If the female finally straightens out in front of the male for long enough, a particularly steep mount can signal intromission (fig. 5.7). Then mating is over for the time being. The partners separate, each going off in a different direction. The female is likely to pause with arched back.

Figure 5.6. Female tries to escape courting male by running into a group as courting male follows (photo by Elizabeth Cary Mungall, courtesy of Kyle Wildlife, Texas, USA).

Figure 5.7. Steep mount for mating as female holds her tail to the side (photo by Elizabeth Cary Mungall, courtesy of Kyle Wildlife, Texas, USA).

Figure 5.8. Curious fawn draws male's attention as it intrudes on courtship of its mother (photo by Elizabeth Cary Mungall, courtesy of Kyle Wildlife, Texas, USA).

When an experienced herd male first starts taking notice of a young female, mating is unlikely. Faced with these new attentions, she runs away. (For more on courtship of young females, see chapter 7). Ordinarily, an adult male ignores fawns unless one gets in his way. This can happen when a buck courts a fawn's mother (fig. 5.8). Then the fawn is apt to be taught that failing to withdraw from a buck's notice can bring on an attack. Learned well when young, this lesson can stand the youngster in good stead because an adult buck is likely to press his attack toward an adolescent or subadult male more seriously.

Birth and beginnings

Sometime less than seven to seven and a half months after successful mating (a time interval that combines recovery from the last birth and the period of gestation), the female gives birth. (For a discussion of reproductive statistics, see chapter 7.) She may separate from the group at least somewhat, or she may merely stay where she is and let the group move on. Consequently, she may or may not be away from other dama gazelles as she goes into labor. If another female comes too close, she is likely to warn it off with a full "head-flagging" dominance display, the same kind used by grown males challenging each other, or by either sex with lesser intensity in various other situations.

Other than the initial labor, which can take an hour or more, the first obvious sign that birth is progressing is the appearance of a balloon-like sac of amniotic fluid protruding from the vulva (fig. 5.9*A*). At this stage, adolescent males may crowd around a young female that is going into labor for the first time. If she is unable to sweep them away with her horns, she may run and be chased. Establishing a peaceful place for birth can be difficult if the female has yet to develop the size and dominance to assert herself against young males.

Table 5.1 gives an example of timing for the birth sequence and the initiation of "lying-out" behavior. The time from appearance of the amniotic sac until

Figure 5.9. Birth to first interactions of mother with newborn: *A*, appearance of sac of amniotic fluid signals that fawn will soon emerge; *B*, first attempt to stand is often just the hind quarters getting up; *C*, neonate starts searching for udder by nosing into whatever angle of the mother is nearest; *D*, mother's check to rear of fawn helps it find udder; *E*, mother cleans umbilical stump under fawn as well as sheath area if fawn is a male; *F*, mother grabs emerging afterbirth; *G*, afterbirth comes free as mother continues to eat it; *H*, curious youngsters crowd around the new fawn; *I*, mother chases away a grown daughter that got too close (photos by Elizabeth Cary Mungall, courtesy of Kyle Wildlife, Texas, USA).

Table 5.1. Timing of the birth sequence and initiation of mother-offspring relations

EVENT	TIME FROM START
Birth sac appears	0 min
Head and front hoofs come out	31 to 40 min
Fawn hangs by hindquarters	51 min
Neonate totally expelled	54 min (expect 45 min to 1 hr 30 min from time 0)
Mother lowers head to neonate	55 min
Neonate shows ear or head movement	1 hr 34 min to 1 hr 43 min
Neonate tries to rise (hindquarters up)	1 hr 35 min to 1 hr 44 min
Neonate first stands on all four legs	1 hr 40 min to 2 hr 3 min
Neonate first searches for teat	1 hr 41 min to 2 hr 45 min
First successful nursing	2 hr 7 min to 2 hr 53 min
Mother first cleans underneath neonate	2 hr 18 min to 2 hr 46 min
Afterbirth appears	2 hr 12 min
Afterbirth all consumed	2 hr 20 min (expect 8 to 9 min to eat afterbirth)
Mother finally separates from neonate	5 hr 41 min to 6 hr 36 min

Source: Composite of birth observations on three different females in the same pasture, courtesy of Kyle Wildlife, Texas, USA.
Note: Female localizing at birth site and lying down are taken as start of initial labor, which can last for an hour or more.

the newborn is on the ground varies but is likely to be 45 to 90 minutes. As for the initial labor, the mother-to-be alternates between standing and lying. The end point of birth often occurs while the mother is standing, letting gravity help pull the fawn for its final drop to the ground. Next comes a period of waiting. It can be 30 minutes or longer before the mother's solicitous licking at this new lump on the ground produces twitching ears or the bob of a head strong enough to be seen above the grass. Very soon, the neonate's haunches will probably waver into view (fig. 5.9*B*). Once both pairs of legs hold at the same time, tottering to the mother and nosing into angles on her body finally takes the neonate to the udder (fig. 5.9*C*). A mother's continued licking tends to push her fawn toward her rear (fig. 5.9*D*). If she lifts a hind leg as her fawn nears its goal, that also helps the search.

The early suckling sessions, which can last for as much as three or four minutes, can seem amazingly long compared to the 15 to 30 seconds that a growing fawn may get. However, part of the first nursing sessions is spent losing and regaining a teat. Soon, the newborn flops back to the ground. As this repeats multiple times, the fawn starts tottering a few steps ahead in between nursing. The mother starts licking the umbilical stump as well as cleaning under the tail and belly as necessary (fig. 5.9*E*). The mother will stand for her newborn offspring to nurse for as long as it wants or as long as it is able. At this early stage, she breaks off nursing only if a disturbance needs her attention.

By the time the afterbirth appears, as much as two hours or more after the birth sac first appears, the mother and fawn may be several yards from the birth site. The mother reaches back repeatedly and grabs the afterbirth with her mouth, pulling at it and eating the whole mass (fig. 5.9*F*). This appears to help extract the afterbirth and can keep it from touching the ground (fig. 5.9*G*). If a part does fall, the mother lowers her nose to it and consumes it, too.

The neonate continues to rise, establish nursing, wobble aside, and collapse onto the ground. If this activity attracts attention, other gazelles—usually curious adolescents—may come to inspect the newcomer (fig. 5.9*H*). If they find it up and stay by it for any extended period, one or another of them is likely to butt the neonate and knock it over. Usually, the mother butts away the intruding youngsters before this happens. The mother is likely to lunge at any approaching adult female (fig. 5.9*I*) and will lead her fawn away from any approaching male. Eventually, the female steps ahead and away from the continued attention of her fawn. Later, the fawn will develop many behavior patterns, but soon after birth, all it seems to do is struggle up when the mother is there, seek out nursing, and then go a few steps before collapsing. Up, nurse, down. Up, nurse, down. Up, nurse, down, as long as the female is there. Finally separating herself, the new mother lies down for a rest. Now, the fawn stays lying down for its first "lying-out" period.

Development of mother-offspring relations

The new mother stays vigilant. If a large bird such as a vulture swoops into the area, she will lunge at it and snort. If an animal on the ground gets close, even one as dissimilar to a gazelle as the two-legged bird in figure 5.10, a rhea (*Rhea americana*), she will give chase and thrash it with her horns if she can get close enough before it clears the area. Small mammals like a squirrel (*Sciurus niger*) have been found flattened near where a newborn had been lying.

When the mother goes back to her fawn, it resumes the up-nurse-down behavior, with increasing time for steps between the "nurse" and the "down." Thus begins the lengthening of both the maternal care sessions and the fawn's resting periods. After nursing and before the fawn goes off and chooses its resting place, the mother cleans it by licking under the tail, and also under the belly for any neonate with an umbilical that is still raw or for a male fawn (fig. 5.9*E*).

As the fawn gains coordination and strength, it adds new behavior patterns. Besides butting into the mother's udder and rapidly wagging its tail during nursing, the dama gazelle fawn may add occasional kicks with a hind leg (fig. 5.11). These kicks may be analogous to the foreleg kicks of a white-tailed deer fawn (*Odocoileus virginianus*) or a domestic horse colt during nursing—or perhaps they just help the dama gazelle fawn keep itself pushed forward into the udder. Dama gazelles stop exhibiting this behavior after about four and a half months. After nursing, the growing fawn adds running steps and hops. These turn into solitary play bouts as the

Figure 5.10. New mother thrashes at a rhea as she chases it away from her fawn (photo by Elizabeth Cary Mungall, courtesy of Kyle Wildlife, Texas, USA).

Figure 5.11. Fawn kicks with hind leg while nursing (photo by Elizabeth Cary Mungall, courtesy of Kyle Wildlife, Texas, USA).

Figure 5.12. Fawn runs in solitary play (photo by Christian Mungall, courtesy of Kyle Wildlife, Texas, USA).

Figure 5.13. Fawn gives a series of stotting jumps in play (photo by Elizabeth Cary Mungall, courtesy of Kyle Wildlife, Texas, USA).

fawn dashes away from its mother and then back again. The running can be quite fast (fig. 5.12), fast enough to put the fawn at risk if it hits a fence. The hops turn into "stotting" jumps (fig. 5.13). If there is a disturbance while the fawn is up, the mother hurries past her fawn, stimulating it to follow her away. The fawn has already mastered both the slow and fast versions of the parallel pacing gait that is the dama gazelle's characteristic mode of locomotion. A diagonal walk comes to be reserved for the slowest progression and a diagonal trot for some transitions. A few trotting steps when disturbed often become a prelude to a run if the disturbance increases.

After one to two weeks, the fawn begins showing interest in other fawns. Fawns of similar age start drawing together after nursing, poking at plants as they taste their world, inciting each other to mutual running games, and lying down together. A rough schedule of nursing every six hours is established: dawn when the adults rise, midday after the adults finish their coordinated morning rest, evening before the adults bed down, and partway through the night while the adults are active again.

By now, it is the mother that ends nursing. Instead of standing until her fawn stops, she lifts a hind leg after a few seconds and steps over and away from her fawn. As she leaves or as a fawn tries to reestablish nursing, she often inflates her nose to give a warning snort (fig. 5.14). The fawn runs along beside her, but she runs faster. When she slows, her offspring may circle in front of her to bar her way (fig. 5.15). Occasionally, this lets a youngster get briefly back into

Figure 5.14. Mother wrinkles her nose and warns her fawn with a snort as she breaks off nursing (photo by Elizabeth Cary Mungall, courtesy of Kyle Wildlife, Texas, USA).

A

B

Figure 5.15. Adolescent tries to reestablish nursing: *A*, blocks mother's way; *B*, tries to get back into reverse parallel position to nurse again (photos by Elizabeth Cary Mungall, courtesy of Kyle Wildlife, Texas, USA).

Figure 5.16. Adolescent objects to end of nursing: *A*, mother cuts off nursing; *B*, adolescent follows mother; *C*, adolescent rubs against mother; *D*, adolescent pushes head against mother (photos by Elizabeth Cary Mungall, courtesy of Kyle Wildlife, Texas, USA).

Figure 5.17. Six-month-old adolescent about to be weaned: *A*, nurses on its knees; *B*, mother ends nursing bout (photos by Elizabeth Cary Mungall, courtesy of Kyle Wildlife, Texas, USA).

the reverse-parallel nursing position, but usually it is unsuccessful. As if in frustration, a youngster may rub its head and neck upward against its mother's shoulder (fig. 5.16). Still with no satisfaction, it may then push its head into her shoulder. This becomes very counterproductive as the growing youngster starts showing horns.

Weaning

With special attention from human keepers and a particular age-mate for companionship, a youngster orphaned as early as three months old can survive. An indulgent mother may let her offspring nurse as long as six months. But by then, the adolescent is eating considerable vegetation, and its defecation has changed from a peanut-butter color and consistency to the dark, firm pellets typical of adults.

Even though the lengthening horns are not an issue because of the nursing animal's head posture, the youngster has grown so much larger that it has difficulty reaching the udder even when down on its knees (fig. 5.17). As this becomes more and more awkward, the mother becomes less and less likely to stand for nursing. Finally, her adolescent is weaned.

Daily Activity

Dama gazelles are early risers. As the first predawn glimmer appears on the horizon, the first of the gazelles rise and stretch. Eventually, each gazelle ambles away with pauses for a few bites here and there. Soon all the adults are up and drifting along in a loose group. If there are mothers in the group, one may separate and work her way apart from the others. The direction she looks indicates the path her young fawn will soon travel as it approaches for nursing. Older fawns often take longer to connect. After these older fawns draw together in twos or threes and move among the adults, these fawns are likely to wait until one of their mothers happens by, when her youngster will sprint to her for a quick suckling bout.

By now, both earth and sky are dimly lit. As perhaps the coolest time of the morning, this is the best daylight chance for leaves to have extra moisture and for dew to form beads on the grass. Understandably, the most concentrated foraging time has begun. Young fawns poke around among the ground plants and make an occasional run out from mother and back. Older fawns do serious eating but also stimulate each other with runs, throwing in a series of stotting jumps at pauses or as a finish to this play. As this flurry of activity dies down, a young fawn will select a spot near the security of tall plants or downed tree limbs and sink down to rest (fig. 5.18). After the sun rises above the horizon, older fawns will lie down also, but they typically pick spots near other older fawns bedded down along the path of the adult group.

Figure 5.18. Fawn nestles by a patch of pricklypear for a rest period (photo by Elizabeth Cary Mungall, courtesy of Kyle Wildlife, Texas, USA).

Meanwhile, dama gazelle bucks have been active, too. Following their first serious feeding of the day, pairs of bucks spar with each other. Other bucks seem attracted to the sparring and advance on the wrestling pair to take over the fight or break it up (fig. 5.19). A buck may thrash his horns against the ground. Off and on, an older buck nearby may check females and try tentative courtship approaches if a female stays within range. If a female proves receptive, then courtship progresses. Bachelor bucks without female company display and spar among each other, but they stay amicably in ones and twos, or threes and fours.

As the sun rises higher and the still of the morning changes to a breeze, the adolescents start to draw together and bed down. This occurs wherever the group happens to be, so these adolescents end up resting as a subgroup, sometimes apart as the older gazelles continue to graze along in their loose association. The adolescents bed down in the open among tall tufts of grass. Older fawns and adolescents, both male and female age-mates that are growing up together, stay together. If an older fawn is already bedded down, then the first adolescents to bed down may cluster around it as they also lie down. As the dew dries off the grass, adults start to bed down, too.

Figure 5.19. Young adult males gravitate together when sparring starts (photo by Elizabeth Cary Mungall, courtesy of Kyle Wildlife, Texas, USA).

A

B

Figure 5.20. Adult male gets a lying female up: *A*, female is reluctant to rise; *B*, female stretches after rising (photos by Elizabeth Cary Mungall, courtesy of Kyle Wildlife, Texas, USA).

By late morning, virtually all the gazelles may be down for their first and best rest period of the day. After an hour or more, first one and then another gazelle rises. An adult male that has already risen is likely to amble through the group getting up first one of the lying animals and then another (fig. 5.20). This helps coordinate the group's activity. Each gazelle stands and stretches. Then the animals move off foraging. The group continues on its way. As rest time ends, females that have small fawns go look for them, and females with a larger fawn or adolescent stand and allow nursing when the youngster discovers its mother close by as the group members graze along.

Foraging continues, although less intensely than in the morning. By late afternoon, youngsters are jostling around together. Immatures of all ages experiment with reproductive (fig. 5.21) and aggressive

A

B

Figure 5.21. Youngsters mount each other during afternoon play: *A*, male fawn to age-mate; *B*, male adolescent to male adolescent (photos by Elizabeth Cary Mungall, courtesy of Kyle Wildlife, Texas, USA).

Figure 5.22. Youngsters practice aggressive behavior: *A*, older fawns spar; *B*, older adolescents spar; *C*, adolescent tries head-flagging to younger herd member (photos by Elizabeth Cary Mungall, courtesy of Kyle Wildlife, Texas, USA).

A

B

C

(fig. 5.22) behavior patterns that they will need later in serious contexts. Just as the sun is going down, play breaks out among the youngsters in the form of running games and stotting. Play is especially likely as temperatures moderate after a hot day. Adults are also busy at this time. Bachelor males resume sparring (fig. 5.23), and adult rivals may indulge in a bout of head-flagging (fig. 5.24). Other gazelles may find a tree and rub their horns along an overhead branch—sometimes the fronts, often the backs (fig. 5.25). Territorial bucks are likely to chase young males in between checking females and testing their receptivity with initial phases of courtship.

Activity subsides as the sky gets dark. Youngsters nurse and then select bedding sites. After a while, only older dama gazelles are on their feet. Then they, too, start lying down. A few here and there lie down in the open until all are bedded. This starts the longest and most uninterrupted resting period of the night.

During the night, dama gazelles still get up periodically, but there is not as much activity, like sparring, running, or courting. Ambling across the pasture and feeding alternate with rest (fig. 5.26). Rest dominates the hours leading up to dawn, and the cycle begins again.

Figure 5.23. Bachelors interlock horns as they spar (photo by Elizabeth Cary Mungall, courtesy of Kyle Wildlife, Texas, USA).

Figure 5.24. Herd male head-flagging in bout with younger adult challenger (photo by Elizabeth Cary Mungall, courtesy of Kyle Wildlife, Texas, USA).

Figure 5.25. Females and young adult male gather to rub horns on denuded branches (photo by Elizabeth Cary Mungall, courtesy of Kyle Wildlife, Texas, USA).

Figure 5.26. Group members have risen after first coordinated rest period of the night and file across pasture at 8:39 p.m., as observed through a night scope (photo by Christian Mungall and Elizabeth Cary Mungall, courtesy of Kyle Wildlife, Texas, USA).

Chapter 6

Territoriality

Elizabeth Cary Mungall

As I have observed in Texas for exotics of the eastern subspecies of dama gazelle (*Nanger dama ruficollis*), the dama gazelle is a territorial species. A segment of the adult male population sets up reproductive territories. Size can vary considerably depending on habitat characteristics and the aggressiveness of particular males, but the maximum can be extremely large.

Typical Behavior

The owner of a territory checks females that come close. Sometimes, this is a female staying apart near her fawn that is still in the lying-out phase. More often, this is one, and then another, of the females in a group of females and young that circulate through his area. If the female is receptive, then the buck's initial approach progresses to more serious courtship. If the female strays near her suitor's border, he tries to block her from leaving his territory. However, a determined female, such as one moving on with her herd, always succeeds in evading the territory owner. Situations like this can precipitate a serious fight if another buck seeks to take over an estrous female.

On days when a receptive female is present, the peaceful pattern of grazing together while drifting along is punctuated with much more male-male activity than usual. The territorial buck enforces the dominance that he enjoys within his boundaries with determined marches after any males that may have come as part of a female group (fig. 6.1). When a pasture has lost its herd male but young males are present, they will pester a female by crowding around

Figure 6.1. Territorial male marches after a young male to send him to the periphery of the master buck's territory (photo by Elizabeth Cary Mungall, courtesy of Kyle Wildlife, Texas, USA).

Figure 6.2. Estrous female pestered by three young males when no herd master is present (photo by Elizabeth Cary Mungall, courtesy of Kyle Wildlife, Texas, USA).

her and trying to mount (fig. 6.2). One function of a territorial male is preventing situations like this by clearing his zone of competition.

Young Bucks and the Start and End of Territoriality

Separated subadult males are likely to team up with other subadult males—often two or three together. Sometimes, a subadult female stays with these subadult males for a while. Occasionally, subadult males visit a territorial male when no estrous female is close by. (For descriptions of age categories such as subadult, see chapter 7.)

A maturing male that is localizing in an area may amble over to a territorial male and indulge in a bout of "head-flagging" with the owner near his border. In this dominance display, first one and then the other of the males draws back his neck into an exaggerated *S* shape, lifts his nose, and turns it (or his horns) briefly toward his opponent (fig. 6.3). The horns-toward gesture may be a swing-out movement for the nose-toward gesture, analogous to the interpretation of a similar behavior of the Grant's gazelle (*Nanger granti*), as if a human were giving a contemptuous toss of the head to signal "get out of here" (Walther 1984). If the two bucks begin circling each other while alternating head-flagging, then the territorial buck may finally change to a medial presentation of horns. In this threat display, he lowers his head to back height with horns pointed at his adversary. Now the intruder must withdraw or fight. With the assurance of ownership on the territorial buck's side,

Figure 6.3. Head-flagging encounter by a young adult male (*left*) challenging an established territorial adult (photo by Elizabeth Cary Mungall, courtesy of Kyle Wildlife, Texas, USA).

the intruder usually withdraws. However, he occasionally accepts the escalating challenge by engaging horns. The bucks push and wrestle and try to slip horns past the other's guard. Each seeks to catch a leg between his horns and flip his adversary (fig. 6.4). A downed buck must scramble up fast. If his opponent does not use the interlude to get away, then the buck that has not fallen will rush in ready to gore with his horns. As with a chased adolescent or subadult male that fails to run fast enough, horn points rake the adversary (fig. 6.5). Most injuries on a male found dead after such an encounter show bruising and cuts on haunches and flanks.

Figure 6.4. Young dama gazelles fighting: *A*, sparring adolescents practice catching each other's legs; *B*, upslope partner gets his adversary low as a different sparring match turns serious; *C*, upslope partner flips his adversary and runs (photos by Elizabeth Cary Mungall, courtesy of Kyle Wildlife, Texas, USA).

A

B

C

Figure 6.5. Herd master thrashes at hindquarters of a maturing son during a vigorous chase (photo by Elizabeth Cary Mungall, courtesy of Kyle Wildlife, Texas, USA).

Unlike some species, such as the Indian blackbuck antelope (*Antilope cervicapra*), in which the males push and wrestle and wait for a fallen adversary to right himself and get back into an appropriate head-to-head position, dama gazelles "fight for damage," as ungulate specialist Fritz R. Walther used to say of similar species. Territorial aggression among dama gazelles may seem a harsh reality, but a buck with a territory—especially a large territory, which dama gazelles keep when there is enough space—expends a large amount of energy in order to keep his mating activities from being thwarted by interference from multiple competitors.

Also unlike blackbuck antelope, dama gazelle males make dung piles that do not seem to be constant points of reference as territorial markers. Instead, dama gazelles are likely to create dung piles at points of current significance such as roads frequently traveled (fig. 6.6), feeders routinely used, and convenient water troughs. If a dung pile on a ranch road gets demolished by road maintenance, it is not necessarily reestablished. Similarly, a dung pile beside a sheltered winter resting place may fall into disuse once the animals change to other sites when winter is over. Nevertheless, a common site for a dung pile is near a spot at which territorial boundary encounters often occur. After such an encounter, a buck is likely to turn back into his territory and exhibit a conspicuous sequence of urination and defecation postures nearby. Thus, a dung pile develops there.

A territorial buck unable to keep up with recurring challenges will eventually lose his place or be killed in the process. Before species numbers declined in Africa in regions where dama gazelles periodically drew together into herds that could number as many as 200 and trekked into the Sahara (John Newby, pers. comm.), it seems unlikely that an aging buck could have retaken a former territory on return to the Sahel. Instead, he might have been forced to withdraw as a bachelor. Thus, territorial periods in such parts of Africa may have been limited to the times between these irregular treks—presumably between irregular rain showers bringing attractive forage in the desert. Additionally, seasonal changes between more open Sahel areas with patches of grass during the rains and the protection of wooded, shaded wadis during the long, hot dry season may also have shortened territorial periods (Jebali 2008, Newby 2015). In contrast, exotic dama gazelles in an exceptionally large West Texas pasture (8,996 ha [22,220 ac])—where space did not appear limiting, where vegetation was fairly uniform, and where

Figure 6.6. Dung pile near split in a road (photo by Elizabeth Cary Mungall, courtesy of Kyle Wildlife, Texas, USA).

changes in seasonal temperatures were not as extreme as in native habitat—did not shift their use areas with season (as studied by Elizabeth Cary Mungall and Susan M. Cooper).

As for territorial periods in native habitat today, with the gazelles now widely scattered and in very low numbers—1, 3, up to perhaps 20 in a couple of really "big" populations (John Newby, pers. comm.) —perhaps a master buck can hold his territorial position throughout his adult life the way a Texas patriarch often can. This assumes that seasonal shifts do not disrupt the pattern. Social conditions as well as environmental conditions can affect the way species-specific behavior is expressed.

Three Texas Examples

In a food-supplemented pasture on the Edwards Plateau, the main part of the familiarly named "Hill Country" of Central Texas, a buck observed throughout his mature lifetime was territorial continuously over the entire span, although territory size finally decreased during his declining years (Mungall, pers. observation). As a new adult of 2 years old, he was given a herd of six adult females. From then through his prime at about 4 through 7 or 8, he killed all maturing sons that were not withdrawn from the pasture. One of his withdrawn sons did the same when given his own herd of four females. Finally, after surviving several serious fights as he aged and as an increasing number of males survived in his pasture, the original patriarch turned up dead two days following a successful but stressful fight at 12 years old. The original master had been territorial in his pasture continuously from 2 to 12 years of age.

In his prime, this vigorous Texas buck (given the designation "DD") held the whole of his pasture of 21.6 ha (53.4 ac) as his territory. As this owner lost space with age, and as newly mature challengers were then able to vie for parts of the pasture, territory sizes of 2.5 to 8.9 ha (6.2 to 21.9 ac) were finally seen (map 6.1). Later, as the new bucks in their adjacent territories became secure within their boundaries after DD died, injuries and then fatalities began to appear.

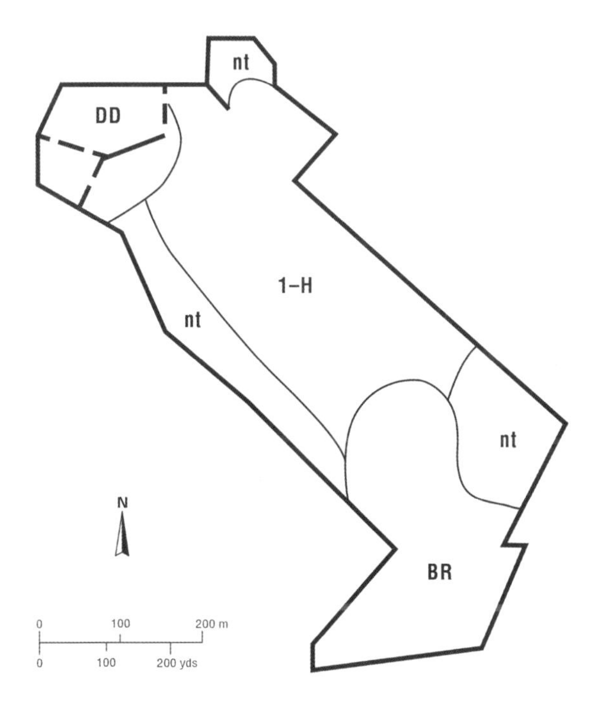

Map 6.1. Territorial division in a Hill Country pasture in Central Texas: the entire pasture including all parts equals 21.6 ha (53.4 ac), and segments are marked with owner's initials or with "nt" for areas not within territories. Time period is when territory size for the senior male, DD, began to decrease.

A comparison with these close observations was made in two other pastures (studies by Mungall and Cooper). GPS-radio collars were used because much of the vegetation was too dense in the additional Hill Country pasture and because distances were too great in the West Texas pasture to observe enough behavior directly. Gazelle distribution within the pastures was calculated as the fixed-volume Kernel Home Range (KHR). This technique is a standard measure of animal distribution based on the mathematical probability of an animal using an area (Seaman and Powell 1996). An area enclosing 50 percent of an animal's recorded locations was taken as the animal's core area, because up to this level the areas where adult bucks localized their activity did not overlap. An area enclosing 95 percent of an animal's recorded locations was taken as the animal's home range. Activity beyond that was considered exploratory behavior.

GPS-radio collar locations for three adult dama gazelle males released into the Hill Country study pasture of 202 ha (500 ac)—where the animals lived on natural vegetation because larger species such as scimitar-horned oryx (*Oryx dammah*) monopolized the feeders—also suggest a flexible range of territory sizes. One buck followed for 11 months established his core area as 36 ha (89 ac), later concentrated to 31 ha (77 ac), of the gently sloping bowl favored by the main female group. During these 11 months, another buck went from a two-part core area of 65 ha (160 ac) to a three-part core area of 80 ha (197 ac). This was across the northern part of the pasture used predominantly by a splinter female group. Core areas for these two bucks ended up with an overlap of only 1.08 ha (2.67 ac). During the last six months of the project, before the collars detached, these two bucks were only recorded within 10 m of each other 21 times, and the median distance between these two males was 884 ± 475 m (967 ± 520 yd).

Meanwhile, the third buck—documented for his first three months—acted more like the subadults in the West Texas study. He utilized a core area (48 ha [118 ac]) with extensive overlap of the core areas of both of the other two adult males that had been added to the pasture. This third male had minimal association with any of the females, though he did have some contact with a maturing male that had been chased out of the female group when the three adult males had arrived. However, this contact was not significant enough to suggest the start of a bachelor group. The two added males that associated significantly with females (with 31 percent and 35 percent of their recorded locations 50 m [55 yd] or less from female company) appeared able to partition the pasture because of topography and vegetation characteristics. With extensive patches of tall brush limiting sight lines and places where rocky slopes reached inclines of 11 to 27° (20 to 50 percent)—gradients that dama gazelles appear to avoid (Mungall and Cooper studies in both Hill Country and West Texas), the two bucks were able to divide the pasture into two separate segments, each with focal use by females.

No fighting was identified during the Hill Country GPS-radio collar investigation, but ranch managers are watching for any problems with the maturing male in case he starts pushing for territorial rights and brings himself into conflict with the established males. If he does, he can be darted and moved to a different pasture on the ranch.

No fighting fatalities were reported among the dama gazelles studied in the huge 8,996 ha (22,220

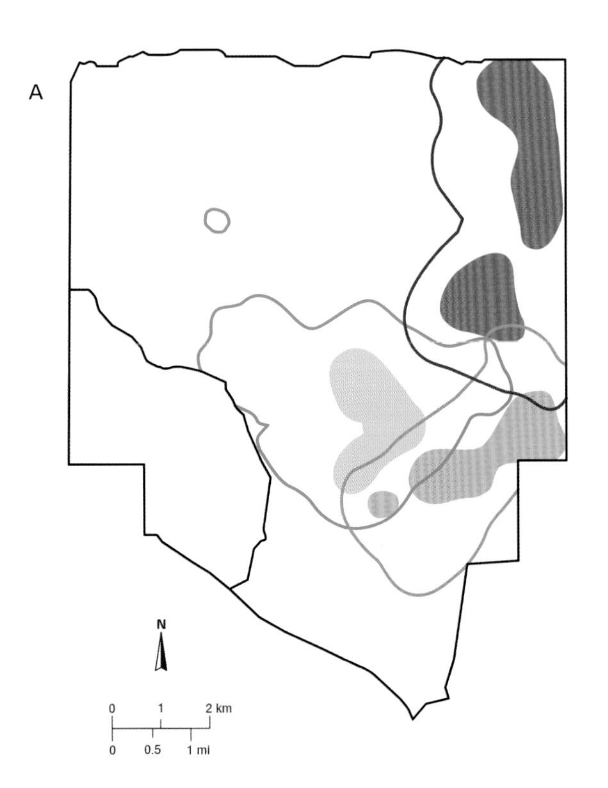

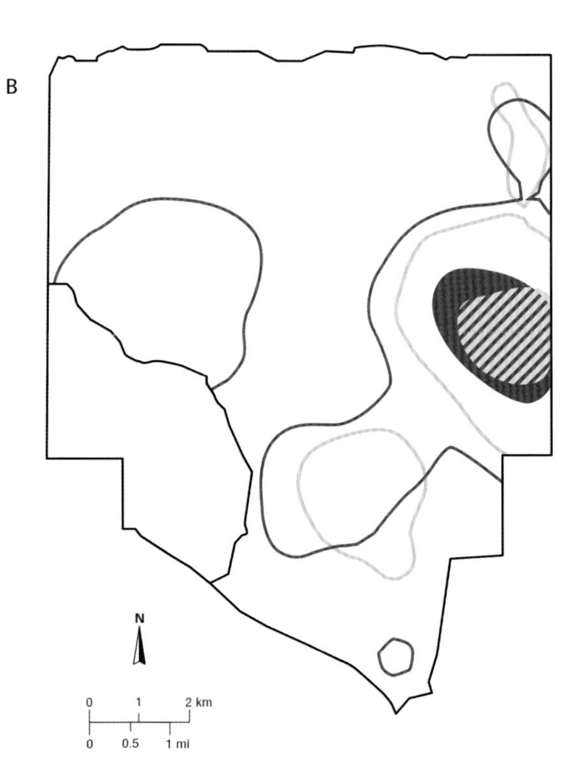

Map 6.2. A year's Kernel Home Ranges (KHRs) for dama gazelle bucks in a West Texas GPS-radio collar study, including 95 percent home ranges and 50 percent core areas of A, three adult males, and B, two subadult males.

ac) pasture in West Texas. There, all gazelles were judged able to move within their habitat without the fences being a limiting factor. GPS-radio collars were worn by three of the adult bucks and two maturing subadults for a year. The data from these collars showed that all three adults maintained nonoverlapping core areas. In the West Texas pasture, the nonoverlapping core areas averaged 440.2 ± 143.7 ha (1,086.8 ± 356 ac). These were within overlapping home ranges (95 percent on the KHR scale) with an average size of 1,783.5 ± 364.4 ha (4,404.0 ± 899.1 ac) throughout the study year (map 6.2*A*).

Even with overlapping home ranges, these adult bucks hardly ever came within 0.8 km (0.5 mi) of each other. Whether the 50 percent KHR use areas in West Texas and in the Hill Country GPS-radio collar study pastures could be considered territories could not be verified without visual data to show territorial boundaries, because subjective boundaries established by the animals themselves are the chief criterion for territoriality (Walther 1972b). Nevertheless, both situations are suggestive. In West Texas, it appears that dama gazelles subsisting on natural forage in semiarid brush country—often with long sight lines—establish very large territories. Core areas (50 percent KHR) for the three collared adult male dama gazelles were 349, 365, and a two-part core totaling 606 ha (863, 902, and 1,497 ac). The dama gazelle's close relative Grant's gazelle also establishes particularly large territories, with a diameter of anywhere up to nearly 2 km (about 1.2 mi) (Walther 1972a). This implies a maximum observed territory size for Grant's gazelle approaching 314 ha (776 ac) in native habitat. Thus, the upper territory size for Grant's gazelle may be somewhat smaller than the size a dama gazelle can hold under suitable living conditions. However, size for both relatives is flexible. An example illustrating the large size for Grant's gazelle is the case of a 60 ha (148 ac) territory measured on the Togoro Plains of East Africa. This single Grant's gazelle territory overlapped almost exactly 13 territories established by the small Thomson's gazelle (*Gazella thomsoni*) in the same place (Walther et al. 1983).

The collared subadult males in the West Texas study frequented larger areas than the adult males. They used core areas averaging 381.6 ± 126.6 ha (942.5 ± 312.9 ac) and home ranges averaging 3,149.7 ± 1,567.3 ha (7,779.8 ± 3,871.2 ac). These core areas and home ranges all overlapped with each other and with the areas used by the adult bucks (map 6.2*A, B*). For both adults and, more frequently, subadults, occasional excursions outside the home range for a few days at a time provide chances to assess conditions farther afield. It is possible that subadults use these trips to explore outside their maternal home range and eventually choose a place in which to become territorial. At the end of the West Texas project, the cuts on the hindquarters of the seemingly more assertive of the GPS-radio collared bucks just reaching adulthood indicated that he may have started testing his possibilities and lost an encounter.

Conclusion

All of this demonstrates that dama gazelles are flexible in their expression of territorial behavior, depending on circumstances. However, it also demonstrates that the consequences of the aggressive component of this behavior include injuries, and even deaths, if the males lack adequate space for the full expression of their behavior. This may be difficult to accommodate in captivity because of the large territory size. Ridges, steep slopes, and stretches of dense brush adequate to break up sight lines do help. Sometimes, younger bucks may use temporal, instead of spatial, separation strategies. Captive populations need to be monitored for growing males that may need to be withdrawn. This can help prevent losses in fenced areas too small or too open for chased individuals to escape from harassment.

Chapter 7

Growing Up and Growing Old

Physical and Social Development

Elizabeth Cary Mungall

Because of their transition from a fawn color during their fawn stage to a different adult coloration as they grow up, dama gazelles seem strikingly different from other gazelles. Actually, the newborns look the same as the young of other gazelles. It is just that dama gazelle fawns quickly start to change.

Along with color changes, many other developments go along with growing up and growing old. The following discussion highlights stages in conformation, body measurements, social relationships, reproduction, and longevity. Aside from general interest, each of these stages has importance for management and conservation planning and for keeping the species alive. Managers use their understanding of development to judge whether progress is normal or whether intervention is needed. Conservationists include developmental criteria when debating what sex, age, and social relationships to use for reintroduction.

Age Categories

You can estimate ages just by looking (figs. 7.1 to 7.8). Color, conformation, and horn shape all give important clues. Table 7.1 charts the changes for the eastern subspecies of dama gazelle (*Nanger dama ruficollis*), which is also called addra, with comparisons to the western subspecies, mhorr gazelle (*Nanger dama mhorr*). Data were collected whenever specimens were available. Thus, numbers of animals representing the different categories vary.

Figure 7.1. Newborn eastern dama gazelle at four days old (photo by Christian Mungall, courtesy of Kyle Wildlife, Texas, USA).

Figure 7.2. Classic fawns: *A*, eastern dama gazelle fawn; *B*, mhorr gazelle fawn (photo *A* by Christian Mungall, courtesy of Kyle Wildlife, Texas, USA; photo *B* by Elizabeth Cary Mungall, courtesy of Safari Enterprises, Texas, USA).

A

B

Figure 7.3. "Pink" fawns: *A*, eastern dama gazelle; *B*, mhorr gazelle (photos by Elizabeth Cary Mungall, *A* courtesy of Kyle Wildlife, *B* courtesy of Safari Enterprises, Texas, USA).

A

B

Figure 7.4. *A*, "Chrome" stage for eastern dama gazelle; *B*, no "chrome" for mhorr gazelle but coat still not reddish like that of adult (photos by Elizabeth Cary Mungall, *A* courtesy of Kyle Wildlife, *B* courtesy of Safari Enterprises, Texas, USA).

A

B

Figure 7.5. Full adolescents show adult coloration: *A*, eastern dama gazelle; *B*, mhorr gazelle (photos by Elizabeth Cary Mungall, *A* courtesy of Kyle Wildlife, *B* courtesy of Safari Enterprises, Texas, USA).

A

B

Figure 7.6. Subadults: *A*, eastern dama gazelle male; *B*, eastern dama gazelle female (photos by Elizabeth Cary Mungall, courtesy of Kyle Wildlife, Texas, USA).

A

Figure 7.7. Adults: *A*, eastern dama gazelle bachelor male at three years old; *B*, eastern dama gazelle bachelor showing maximum neck development; *C*, eastern dama gazelle female; *D*, mhorr gazelle male; *E*, mhorr gazelle female (photos *A* and *C* by Elizabeth Cary Mungall, courtesy of Kyle Wildlife, Texas, USA; photo B by Elizabeth Cary Mungall, courtesy of Stewards of Wildlife Conservation, Texas, USA; photo *D* by T. Abáigar/CSIC, courtesy of Estación Experimental de Zonas Áridas, Almería, Spain; photo *E* by Christian Mungall, courtesy of Safari Enterprises, Texas, USA).

B

C

D

E

A

B

C

D

Figure 7.8. Old adults: *A*, eastern dama gazelle male with coat dirty from eating sticky plants; *B*, eastern dama gazelle female; *C*, mhorr gazelle male; *D*, mhorr gazelle female (photos by Elizabeth Cary Mungall, *A* and *B* courtesy of Kyle Wildlife, *C* and *D* courtesy of Safari Enterprises, Texas, USA).

Table 7.1. Age class characteristics for dama gazelles

AGE CLASS	MALE	BOTH SEXES	FEMALE
FAWN			
Newborn		Birth to 1 wk	
		Weight 3.9 kg (8.7 lb), height 53.2 cm (21.0 in)	
		Weight *N* = 8, height *N* = 6	
		Brown to light brown, somewhat uncoordinated, body short and thin, legs long, neck thin and short, back lower than mother's belly	
Classic fawn		1 wk to 1 mo	
		Weight 6.3 kg (13.8 lb), height 54.1 cm (21.3 in)	
		Weight *N* = 5, height *N* = 3	
		Light brown, coordinated, body filled out, classic fawn proportions. Mhorr dark eye stripe, nose gray, dark between ears.	
"Pink" fawn		1 mo to 2 mos	
		Weight 9.2 kg (20.3 lb), height 58.4 cm (23.0 in)	
		Weight *N* = 3, height *N* = 3	
		Nanger part of rump patch shows as a pinkish zone, has classic fawn proportions but growing taller, face is still tan (rich tan). Mhorr black eye stripe and nose patch, black between ears and above white eye rings.	
ADOLESCENT			
"Chrome"		2 to 4 mos	
		Weight 17.3 kg (38.2 lb), height 68.4 cm (26.9 in)	
		Weight *N* = 5, height *N* = 4	
		Adolescent proportions, changes closer to adult coloration as hair sheds, *Nanger* part of rump patch goes white, addra shows weak extensions of light rump patch prongs extending forward along sides (separating back color from lower sides that will go white), addra haunches lighten except that upper border stays medium tan shade (will later go roan in many addra or chestnut in central dama gazelles to form horizontal part of any haunch stripe) and border of gazelline rump patch stays tan (will later go roan or chestnut in central dama gazelles to form vertical tail of haunch stripe), face starts to lighten, neck patch gains adult shape, horns start to show. Mhorr goes rich tan but not yet reddish.	

(continued)

Table 7.1. Continued

AGE CLASS	MALE	BOTH SEXES	FEMALE
Full adolescent	Weight 31.3 kg (69.0 lb) Height 81.7 cm (32.2 in)	4 to 13 mos	Weight 22.4 kg (49.4 lb) Height 82.6 cm (32.5 in)
	Weight $N = 3$ Height $N = 2$	Weight range 10.9 to 39.0 kg (24.0 to 86.0 lb) Height range 71.1 to 92.2 cm (28.0 to 36.3 in)	Weight $N = 7$ Height $N = 14$
		Adult color except may still be pigmented hair where borders of pigmented haunch area were (haunch color remains in mhorr or, for light-phase mhorr, shrinks only slightly), face goes white or whitish except for variable amount of gray or similar color on forehead and blackish nose patch for mhorr, horns lengthen and then start to curve forward and then grow about 4 to 5 rings as hair held up at horn bases looks like forehead ridges.	
SUBADULT			
	Weight 42.7 kg (88.8 lb) Height 91.4 cm (35.0 in)	13 to 20 mos	Weight 36.8 kg (81.0 lb) Height 92.0 cm (36.2 in)
	Weight $N = 2$ Height $N = 1$		Weight $N = 2$ Height $N = 8$
	Enlarged C-shaped horns with points roughly horizontal and forward, horn shanks lengthening, male horns thicker than female horns.	Body proportions approach those of adult, horn bases look like forehead ridges, variable amount of gray extends down forehead from horn bases, dark mhorr nose patch lightens.	Horns mainly straight and parallel with lengthened lower portion and tips curved inward, female horns thinner than male horns.
ADULT			
Young adult	Weight 53.6 kg (118.0 lb) Height 101.2 cm (39.9 in)	20 to 24 mos	Weight 44.0 kg (97.0 lb) Height 97.4 cm (38.4 in)
	Weight $N = 2$ Height $N = 2$		Weight $N = 1$ Height $N = 1$
	Lower horn curve minimal, tips point in, neck thickens.	Horns gain S shape, impression of ridges on forehead at horn bases minimal.	Lower horn portion looks short, horn tips point in.

(continued)

Table 7.1. Continued

AGE CLASS	MALE	BOTH SEXES	FEMALE
Full adult	Weight 55.8 kg (123 lb) Height 101.6 cm (40.0 in)	2 to 7 or 8 yrs	Weight 40.9 kg (90.0 lb) Height 97.0 cm (38.2 in)*
	Weight *N* = 3 Height *N* = 4		Weight *N* = 3 Height *N* = 0*
	Horn tips come to point mostly forward, both horn curves pronounced, neck enlarges (especially on bachelors), and more folds in front of withers.	Full S of species-specific horn shape, face goes white except mhorr may retain light gray nose patch.	Horn curves usually shallow, horns with about equal parallel lower part and upper bow part.
Old adult	Weight 57.6 kg (127.0 lb) Height 100.0 cm (39.4 in)	7 or 8 to 15+ yrs (pasture) or 19 yrs (zoo maximum)	Weight 36.9 kg (81.2 lb) Height 96.9 cm (38.0 in)
	Weight *N* = 1 Height *N* = 1		Weight *N* = 4 Height *N* = 2
	May hold reasonable body condition, may be more competitive at feeders than old females.	Withers start to protrude, back may slant down between withers and croup even if haunches not hunched, nose elongates, face is white except mhorr may retain light gray nose patch.	Body looks bony, underline slopes up before hind legs so body looks thin at flank level, may still be nursing a fawn, lower horns long.

Note: Average weights and heights are from animals of the eastern subspecies raised as exotics on two Texas Hill Country ranches. Central dama gazelle comments are inferred from addra and mhorr gazelles. Observed weight and height ranges are given for full adolescents because they grow so much over their long time span as adolescents.

*Estimated as midway between values in the previous and following classes.

The older fawns have been labeled "pink" in table 7.1 because the lightening *Nanger* rump patch looks pinkish in contrast to the rest of the coat. The next stage, the beginning of adolescence, is when the youngster has lost its fawn proportions. This stage for addra starts as the coat is in conspicuous transition. It is labeled "chrome" in the table because of all the special elements—like elaborate trim—that mark the height of this transition. Mhorr young lack this as a special stage because they keep their side color and haunch color, going only from a rich tan to a glossy reddish brown (fig. 7.4*B*). The central subspecies of dama gazelle (*Nanger dama dama*) would be expected to have a less pervasive "chrome" stage than addra because its main change is merely the lower extent of the pigmented hair on the haunches. This would leave the haunch stripe. Whether the central dama gazelle would have a less extensive "chrome-type" lightening zone along the sides, albeit lower, would need to be checked.

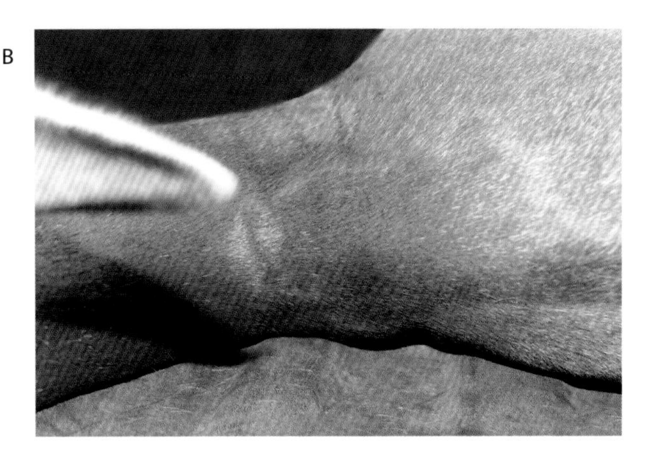

Figure 7.9. Characteristic hair reversal of the dama gazelle: *A*, whorl near withers to line on lower dorsal neck; *B*, whole whorl, reversal, and line are sometimes absent (photos by Elizabeth Cary Mungall, courtesy of Kyle Wildlife, Texas, USA).

The central subspecies of dama gazelle has less intrusion of white into the body color than the eastern form, so that the back and side color stays chestnut instead of turning to roan. As determined by Cano Perez (1984) in her comprehensive evaluation of skins and live animals, the central subspecies has no saddle lightening, and the vertical extension of the haunch stripe extends to the vicinity of the hocks (see appendix 2 for a diagram showing the "hock" and similar designations). Mhorr gazelles in the west retain the most color, are typically darker and redder, and show a dark eye stripe like that of many other gazelle species.

As noted among the ranch addra, no correlation was found between the extent of color and the small number of ranch individuals that lacked the line and whorl of the hair reversal that usually lies between the withers and lower neck (fig. 7.9). The large changes during the adolescent stage, once the basic form of the individual's adult coloration is reached at about four months, are in height and horn development. From the horns just showing above the hair at the start of adolescence, they elongate, curve forward, and finally add their first four to five rings.

Body Measurements and Special Developmental Attributes

Table 7.2 gives body measurements for adults of the eastern dama gazelle subspecies as recorded from exotics in pasture populations on Texas ranches. The number of samples for each characteristic tends to vary because not all live animals and not all carcasses were available for all measurements. For example, some of the live animals were having GPS-radio collars put on and had to be released quickly, and some carcasses had already been broken into by scavengers. Sclater and Thomas (1897–1898) give the height of a male "not fully adult" mhorr gazelle (horns not yet S shaped) as 89.8 cm (35.5 in) and a mhorr gazelle female with S-shaped horns as about 86.4 cm (34.0 in), although the latter was possibly not fully grown. For comparison with the averages in table 7.2, adults of the eastern race as exotics on Texas ranch land average 102.8 cm (40.5 in) for males and 90.7 cm (35.8 in) for females.

Table 7.2. Body measurements for adults of the eastern subspecies of dama gazelle raised as exotics on Texas ranches

ADULT MEASUREMENTS	MALE		FEMALE	
Live weight	55.4 kg, *N* = 6 (range 54.5 to 59.0 kg) 54.5 to 84.9 reported*	122.0 lb, *N* = 6 (range 120.0 to 130.0 lb) 120 to 187 reported*	40.4 kg, *N* = 4 (range 35.9 to 44.0 kg) 40.0 to 65.4 reported*	89.0 lb, *N* = 4 (range 79.0 to 97.0 lb) 88 to 144 reported*
Heart girth (a standard measurement around the body at about the level of the heart)	88.2 cm, *N* = 4 (range 86.8 to 90.1 cm)	34.8 in, *N* = 4 (range 34.2 to 35.5 in)	83.5 cm, *N* = 2 (range 80.7 to 86.3 cm)	32.9 in, *N* = 2 (range 31.8 to 34.0 in)
Shoulder height	102.0 cm, *N* = 4 (range 99.8 to 102.8 cm)	40.2 in, *N* = 4 (range 39.3 to 40.5 in)	97.5 cm, *N* = 2 (range 97.4 to 97.6 cm)	38.4 in, *N* = 2 (range 38.4 to 38.4 in)
Foreleg length (to hoof tips)	74.4 cm, *N* = 4 (range 72.0 to 78.0 cm)	29.3 in, *N* = 4 (range 28.4 to 30.7 in)	72.3 cm, *N* = 2 (range 71.4 to 73.3 cm)	28.5 in, *N* = 2 (range 28.1 to 28.8 in)
Hind leg length (from hock)	50.0 cm, *N* = 4 (range 49.0 to 52.3 cm)	19.7 in, *N* = 4 (range 19.3 to 20.6 in)	46.8 cm, *N* = 2 (range 46.4 to 47.2 cm)	18.4 in, *N* = 2 (range 18.3 to 18.6 in)
Total length (with tail)	170.6 cm, *N* = 3 (range 170.2 to 171.0 cm)	67.2 in, *N* = 3 (range 67.0 to 67.4 in)	163.3 cm, *N* = 2 (range 159.0 to 167.6 cm)	64.3 in, *N* = 2 (range 62.6 to 66.0 in)
Ear length (from notch)	15.3 cm, *N* = 4 (range 14.8 to 16.4 cm)	6.0 in, *N* = 4 (range 5.8 to 6.5 in)	16.0 cm, *N* = 2 (range 15.8 to 16.2 cm)	6.3 in, *N* = 2 (range 6.2 to 6.4 in)
Neck circumference at neck patch	40.6 cm, *N* = 9 (range 37.2 to 48.2 cm)	16.0 in, *N* = 9 (range 14.6 to 19.0 in)	30.3 cm, *N* = 10 (range 26.0 to 33.1 cm)	11.9 in, *N* = 10 (range 10.2 to 13.0 in)
Neck circumference at base (perpendicular to axis of neck)	46.6 cm, *N* = 6 (range 41.5 to 49.3 cm)	18.4 in, *N* = 6 (range 16.4 to 19.4 in)	37.6 cm, *N* = 10 (range 35.5 to 41.6 cm)	14.8 in, *N* = 10 (range 14.0 to 16.4 in)
Horn length (base to tip along front curves)	32.1 cm, *N* = 10 (range 27.7 to 37.0 cm)	12.6 in, *N* = 10 (range 10.9 to 14.6 in)	27.6 cm, *N* = 9 (range 22.5 to 32.0 cm)	10.9 in, *N* = 9 (range 9.5 to 13.0 in)
Horn length (along front and then along back if there is a point of inflection—S curve)	35.7 cm, *N* = 11 (range 31.5 to 40.4 cm)	14.1 in, *N* = 11 (range 12.4 to 15.9 in)	28.8 cm, *N* = 9 (range 24.1 to 33.0 cm)	11.3 in, *N* = 9 (range 9.5 to 13.0 in)
Horn spread (tip to tip)	18.4 cm, *N* = 4 (range 11.3 to 27.6 cm)	7.2 in, *N* = 4 (range 4.4 to 10.9 in)	10.3 cm, *N* = 8 (range 1.0 to 21.2 cm)	4.6 in, *N* = 8 (range 0.4 to 8.4 in)

Note: Data are from three different ranches.

*Reports from literature summarized in Mungall (2007a), including typical male weight of 62.6 kg (138 lb).

A

B

Figure 7.10. Fawns and young adolescents include a wide range of contortions as they jump: *A*, fawn; *B*, young adolescent (photos by Elizabeth Cary Mungall, courtesy of Kyle Wildlife, Texas, USA).

Sounds, especially various levels of snorts, are important for dama gazelle communication. As reported for a hand-raised fawn being tracked through her early developmental stages (for more on hand raising, see chapter 10), after her first day, the fawn was very quiet, but on day 3 she became quite attentive when people made sounds to her. At two weeks of age as the fawn became more active, she started giving snorts when wanting to be put down rather than carried or when wanting to have more milk from her bottle. At six weeks, vocalizations increased. Snorts to her caregiver could be very soft, or very loud when agitated. When given a young goat for companionship, the three-month-old study fawn would answer the goat's bleats with dama gazelle snorts. Similarly, the goat often responded to a snort with a bleat. As illustrated in chapter 5, the mother often gives a warning snort to her fawn when breaking off or discouraging nursing. Ranch dama gazelles that are accustomed to supplemental feeding will also snort as they gather around in expectation of a ration of pellets from a nearby human. A dama gazelle male courting in a low-stretch posture with ears down and out to the sides will sometimes add a series of grunts as he follows his partner.

As far as physical development, it is interesting to note the prevalence of the strange leaps that fawns indulge in as they play (fig. 7.10). These are not used to go over any obstacle, and they can appear spontaneously from a youngster acting by itself or from two or more fawns jostling with each other. In the latter case, the fawns often seem to incite each other as they face one another and experiment with various jumping attitudes. This is seen especially among fawns but also occurs among young adolescents. After that, the dama gazelles perform stotting jumps in play, in excitement, and when prompted by rapidly improving environmental conditions, such as moderating temperatures at the end of a hot day or cessation of rain when the sun comes out after a downpour. Whether performed by fawns or, as happens less frequently, by adults, stotting is very predictable in its details (fig. 7.11). Stotting has none of the flexibility of form that characterizes the leaping contortions of the little youngsters. Temple Grandin has offered an insight relevant to what may be happening with this free-form jumping of fawns and adolescents: the solo running, jumping, and spinning of locomotor play during key phases of physical development of young animals may facili-

A

B

Figure 7.11. Stotting: *A*, an adult male; *B*, an adult female (photos by Elizabeth Cary Mungall, courtesy of Kyle Wildlife, Texas, USA).

tate development of the cerebellum (Grandin and Johnson 2006). This is the region of the brain that controls posture, balance, and coordination.

Social Organization

Social organization changes with age just as height, color, and conformation do. For the first one to two weeks, the fawn's social world revolves around its mother (see details in chapter 5). After that, youngsters are curious about each other, and juveniles and adolescents of similar age team up for almost everything except nursing. They would probably switch off and nurse from any nearby mother if the mother did not butt away any except her own young. She may have two to five teats (ordinarily two are used for nursing), but she reserves them for her own offspring.

After adolescence, males and females segregate for much of the time, although a subadult female may temporarily join a couple of subadult males that are out on their own. Even a subadult female that grows up in close association with male age-mates prefers to go off with the older females instead. The increasing sparring among males and the new attention from both rowdy subadult males and experienced adult bucks are driving forces for the separation. This is not to say that subadult females never use their well-shaped horns in their own sparring matches, however. Occasionally, a subadult female will engage an adult female instead of keeping her distance as usual (fig. 7.12). Very occasionally, a higher-ranking subadult female will pester a lower-ranking subadult female until she accepts an engagement. Such a match can be lengthy, with the subordinate trying periodically to break off the encounter but the dominant partner not letting her go. One such encounter lasted 13 minutes. Other rank relationships can also show up among maturing females. It can happen that a subadult female will use much of the same sequence toward a subordinate female age-mate as would be expected when a male is courting a female—following, standing behind, seeming to push forward in an erect posture, getting her subordinate partner to move ahead, and then emphasizing her dominance with low mounts (fig. 7.13).

Figure 7.12. Sparring between females: the subadult, *right*, is pushing against an adult (photo by Elizabeth Cary Mungall, courtesy of Kyle Wildlife, Texas, USA).

A

B

Figure 7.13. Dominant subabult female "courts" her subordinate half sister: *A*, standing erect behind; *B*, steep mounting (photos by Elizabeth Cary Mungall, courtesy of Kyle Wildlife, Texas, USA).

As a young male grows larger and heavier nearing his subadult stage, he approaches the adult females in size and strength. A female may still accept a presentation of horns from one of these males and engage horns, but she is less likely than formerly to sustain the encounter or to send the young male off with a shove as she would a smaller adolescent. Instead, after locking horns briefly, she may depart as she kicks up her heels and lowers her head in submission (fig. 7.14). Given the greater size and weight of grown males, all females are subordinate to all adult males.

An adult herd master will drive off subadult males and may even chase away adolescent males. Thus, the period from six months, when weaning is complete, to about one or even approaching two years of age, while a male is a subadult, can be a particularly stressful time. Where there is room to reassort, males probably separate from their female groups at about eight months of age (Mungall 2007b). In a really large pasture where there is a large population of eastern dama gazelles, like the native rangeland in West Texas that measures more than 8,996 ha (more than 22,220 ac), an immature male that has been driven off may stay away and live alone in the brush for days (fig. 7.15) before seeking companionship with other dama gazelles again. Harassment by dominant adult bucks keeps boisterous young males from bothering females when they come into heat. When a young adult male dares to engage an older herd master in head-flagging dominance displays, the young adult risks

A

B

C

Figure 7.14. Encounter between a young male and an adult female: *A*, young male presents horns to grown female; *B*, female accepts the challenge; *C*, female kicks up her heels and bounds away in submissive posture (photos by Elizabeth Cary Mungall, courtesy of Kyle Wildlife, Texas, USA).

Figure 7.15. Adolescent male living alone in the brush (photo by Elizabeth Cary Mungall, courtesy of Stevens Forest Ranch, Texas, USA).

Figure 7.16. Subadult male forces the herd master to spar by poking him in the neck when, initially, he did not accept the challenge (photo by Elizabeth Cary Mungall, courtesy of Kyle Wildlife, Texas, USA).

the exchange turning into a brief but violent fight. In this way, an aging master buck may be deposed or even killed. When dama gazelle bucks fight seriously, they "fight for damage," as gazelle specialist Fritz R. Walther says (see chapter 6).

In spite of this, a grown male is often surprisingly tolerant when approached by a subadult male that is trying to engage him in a sparring match. The subadult presents horns near the adult's head, but the big male just stands. There are limits, however. If the subadult finally pokes his horns into his senior (fig. 7.16), then a quick pushing match ensues. The subadult is driven backward decisively, and the encounter is over (Mungall 2007b).

Figure 7.17. Half brothers born only one day apart; the younger male in front is growing faster and becoming more assertive after being raised by a particularly dominant and solicitous mother (photo by Christian Mungall, courtesy of Kyle Wildlife, Texas, USA).

Relationships established between young male age-mates can have serious consequences. It has been repeatedly observed that the dominant of a pair of subadult males matures faster, is more precocious in his behavior, and draws more harassment from a territorial buck (fig. 7.17). This can get the dominant youngster killed by a pasture's territorial buck if there is no place to escape. Interestingly, the psychological inhibition imposed by the superior companion lasts only as long as the subordinate is with his superior age-mate. If released from the domineering effects of the other's presence, he develops into a competitive buck in his own right and can become a strong territorial buck himself. In a pasture of limited size, he in turn will kill even his own growing sons if they are not withdrawn.

While all of these interactions punctuate the day's activity periods, adult and subadult females circulate through their ranges. Usually they travel in small groups of up to half a dozen or a dozen, pausing to take care of their fawns. Except when they pull together in response to a disturbance, the animals are often loosely spread out across the pasture. (For more on the characteristic pattern of daily activity, especially as it involves fawns and adolescents, see chapter 5.)

Reproduction

When a young female begins to attract courtship from a grown male, this is a new experience. As shown in the sequence of photos in figure 7.18, an adolescent's first reaction is to go to her mother. However, her mother moves away, just as she would if the male were courting her, abandoning her daughter to the attentions of the male. The youngster goes into a submissive posture as she moves on. When the male keeps up his approach, she next stops and draws herself up in resistance. When this does not send the male away either, she wheels as she lowers her head in submission, kicks up her heels, and flees. This is the same escape maneuver shown by the grown female in figure 7.14*C*, even down to each of the females having ears lowered to the side. In both cases, the male lets the female go—for the moment. Before a young female begins to cycle, there can be low-level aggression from an adult master of a breeding group, but this does not happen very often (Speeg et al. 2014).

A male that has charge of his own herd sometimes develops a favorite among the young females. He may approach her more often and at a younger age—even as an adolescent, as in figure 7.18. Other growing females are more likely to be left until they become subadults before he begins his advances. Only with experience does the young female adopt the more relaxed responses of the mature female. Thus, a major period of stress in the life of a growing female is when she is six months to a year old, especially toward the beginning of this period (Mungall 2007b). (For details on typical, mature courtship, see chapter 5.)

As tracked in Texas, competition among grown bucks seems to keep the males more attentive to their duties as herd sires. The main Central Texas research herd started out with just a small group size (one male and six females) and had births only in August and September. The birth pattern did not spread out markedly through the year until a second breeding herd was established across a double fence line through which the adjacent territorial males could easily see and hear each other. After that, the only year in which births again occurred only in August and September was an extreme drought year in which

A

B

C

Figure 7.18. Adolescent female first exposed to courtship by experienced male: *A*, male targets adolescent; *B*, adolescent stops and raises her head in resistance; *C*, adolescent finally lowers her head in submission as she kicks up her heels and runs away (photos by Elizabeth Cary Mungall, courtesy of Kyle Wildlife, Texas, USA).

summer temperatures were higher than normal for longer than normal.

Wildlife managers always find it important to know when during the year to expect fawns. Based on data from facilities in the Species360 system (formerly the International Species Information System), the major birth season for the eastern subspecies of dama gazelle as kept in the Northern Hemisphere is from late July to early September (Zerbe et al. 2012). As well as in Texas, this is true in Florida, as tracked for many years at a conservation breeding facility (Siegel 2013, Speeg et al. 2014). Although fawns can come at any time—and Texas can have a secondary peak in March or April—few births occur in January in either region. In northeastern Florida at White Oak Conservation Holdings as well as in Central Texas at the Fossil Rim Wildlife Center, births peak in August following maximum average monthly temperatures in July (Speeg et al. 2014). Relationships with rainfall patterns are less clear. A definite peak in precipitation follows the birth peak in Florida at White Oak, whereas the precipitation peak following the birth peak in Texas at Fossil Rim is significantly lower than the major rainfall maximum that comes in May and June, when temperatures are more moderate and the incidence of births is low (Speeg et al. 2014). For wild populations in Africa, synchronized births are reported from July to September following the start of the rainy season (Sahara Conservation Fund 2013, cited in Speeg et al. 2014).

Monitoring of progestogen and estrogen levels in captive US addra females has not only verified estrus duration and gestation length but has also led to the discovery of a definite propensity toward seasonal anestrus (Wojtusik et al. 2017). Anestrus was defined in this study as more than one month with progestogen levels continuously at or below the established baseline. This means that estrous cycles have ceased and no conceptions would be possible during the anestrous period. Using eight females (1 to 9 years old) for varying intervals during the seven-year study, the average time from the start of one estrous cycle to the start of the next was 19.5 ± 0.4 days (range 14 to 26 days, $N = 61$ cycles). This coincides closely with 18.6 ± 0.3 days as determined for the mhorr subspecies of dama gazelle (Pickard et al. 2001).

Measured for six pregnancies (four different addra females), mean gestation length was 6.6 months (200 to 202 days). Some, but not all, of the captive US females ceased cycling from September to March. However, the female that continued to cycle all year in 2011–2012 had shown the seasonal anestrous state when tested during a previous period and also showed seasonal anestrus when tested during two subsequent periods. For most of the year, estrous cycling typically resumed within a month after parturition. Lactation did not show a depressing effect. For births during the US birth peak (late July to early September), an anestrous state followed parturition.

Day length, rainfall, and temperature were all considered as possible triggers for the timing of reproductive seasonality in both the US (study animals at the Smithsonian's National Zoological Park in Washington, DC; at the Smithsonian Conservation Biology Institute in Front Royal, Virginia; and at White Oak Conservation Holdings near Yulee, Florida) and remaining native habitat (Ouadi Rimé–Ouadi Achim in Chad, Tamesna in Mali, and Termit and Tin Toumma in Niger). Day length seems a prime contender, even surviving translocation of these gazelles to become exotics in the United States. In all the locations evaluated, the September or October beginning of anestrus coincided with day length starting to decrease below 12 hours. Increase back to 12 hours of daylight in March correlated with resumption of cycling. Although annual photoperiod fluctuations in these parts of Africa are less than in the US regions studied, timing by this critical 12-hour threshold would favor births during peak rainfall months in both the United States and in the native Sahelo-Saharan region. Here, monthly precipitation is minimal from October to May (1909 to 2012 range 0.0 to 0.7 mm, 0.0 to 0.03 in), climbs in June (5.6 mm, 0.2 in), and is at a maximum from July to September (range 14.4 to 73.3 mm, 0.6 to 2.9 in). Increased rainfall would be expected to give females their best chance to provide adequate milk for nursing fawns. The lowest percentage of births in the United States came during the time of least rainfall (December to February) and the lowest temperatures for the United States and Africa. Nevertheless, correlations between temperature and numbers of births per month as reported by zoos and similar institutions in the North American addra studbook (Their 2015) were weaker than the relationships with photoperiod or rainfall.

This work on addra by Wojtusik et al. (2017) has clear implications for managers involved with this species. Manipulations for breeding purposes, and especially any artificial insemination work to add genetic diversity to a herd or for other considerations, need to take into account the cyclic nature of dama gazelle reproductive mechanisms.

Another aspect of fawning that can concern managers is whether the herd is producing more of one sex year after year instead of the expected 1:1 sex ratio. When tallied for the main Texas ranch research herd of exotic eastern dama gazelles, the ratio was close to 1:1. However, when the preserve in northeastern Florida had a long run of males, it decided to have its records examined. Only 2 years of the last 12 had produced a fairly even sex ratio of births. In only 1 of those 12 years had female births predominated over males. Looking through data from 1980 to 2010, starting almost from when the preserve began working with the species, Siegel (2013) found that the sex ratio at birth had been 50.4 percent females to 49.6 percent males—almost exactly 1:1. There had been 11 years with more males born, 11 years with more females born, and 8 years with about equal numbers (fig. 7.19). This was then compared with the findings in a paper on bias trends in many species of hoofed animals, carnivores, and primates (Faust and Thompson 2000). The authors reported that 49 of 53 of the species evaluated had a run of same-sex births lasting 7 or more years. Similarly, 24 of these 53 species had a run of 10 or more years. As Siegel concluded, managers involved in planning need to consider the possibility of long sequences like this and to be prepared to compensate if necessary. Particularly when a small population is involved, adjustments may be indicated in order to keep the herd going.

One important aspect for the growth of fawns is nursing. Dama gazelles grow very fast with very short nursing sessions. There are other animals that also have very short nursing bouts, but the contrast between these species and species like nyala (*Tragelaphus angasi*) and domestic cattle can be extreme. For example, a month-old dama gazelle kept under observation during a 16-hour dawn-to-dusk period received hardly more than a cumulative total of 3 minutes of nursing. During that same dawn-to-dark watch, a month-old nyala in a herd across the fence line suckled for nearly 11 minutes in just one of its nursing bouts (Mungall 2010). Table 7.3 compares the composition of a dama gazelle milk sample taken at three months in the lactation cycle (Mungall 2010) with the composition of samples from various other ungulates.

Although composition shifts during the course of lactation—especially in the first three to four days postpartum from the colostrum of the first day, as studied in eland (*Taurotragus oryx*) (Treus and Kravchenko 1968)—the ranges in the table taken over entire courses of lactation, or at least sampled early and late in lactation, show that these shifts in major components are not necessarily dramatic. Therefore, even a single analysis should give a reasonable indication of percentages for the species. Of course, this assumes that other sources of variation, such as differences among mothers, or sampling at different stages of depletion of the udder, are not a problem for the comparison. Among the values for 23 ungulates, both wild and domestic, the closest match, other than for lactose, was to the other gazelles sampled, to the gazelle-like impala (*Aepyceros melampus*)—and to the sika deer (*Cervus nippon*). Similarity of milk composition is considered to follow similarity of habits and habitats more than

Figure 7.19. Example of same-sex runs in sex ratios as seen for a population of exotic dama gazelles in Florida (courtesy of Amanda Siegel 2013).

Table 7.3. Composition of dama gazelle milk compared to that of a selection of other hoofed species

SPECIES (reference)	NO. ANIMALS	NO. SAMPLES	FAT (%)	PROTEIN (%)	LACTOSE (%)	TOTAL SOLIDS (%)
Dama gazelle (Mungall 2010b)	1	1	20.1	12.0	1.2	36.1
Grant's gazelle (Ben Shaul 1962)	—	—	19.5	10.4	2.8	34.1
Thomson's gazelle (Ben Shaul 1962)	—	—	19.6	10.5	2.7	34.2
Mountain gazelle (Ben Shaul 1962)	—	—	19.0	12.4	3.3	36.1
Impala (Ben Shaul 1962)	—	—	20.4	10.8	2.4	35.3
Springbok (van Zyl and Wehmeyer 1970)	1	26	8.9 (5.2 to 13.3)	8.0 (5.5 to 11.5)	4.8 (3.9 to 5.2)	21.9 (17.1 to 27.2)
Blackbuck antelope (Dill et al. 1972)	3	3	9.3 (7.8 to 11.2)	6.9[a] (6.8 to 7.0)	4.3 (3.6 to 4.8)	20.8 (19.5 to 22.4)
Pronghorn (Einarsen 1948)	1	1	13.0	6.9	4.0	24.9
Eland (Treus and Kravchenko 1968)	> 9?	51	9.8 (8.2 to 10.9)[b]	12.7 (11.7 to 13.4)[b]	3.9 (3.5 to 4.4)[b]	21.8 (19.7 to 23.3)[b]
Greater kudu (Vice and Olin 1967)	1	1	10.0	12.2	4.6	36.1
Sika deer (Ben Shaul 1962)	—	—	19.0	12.4	3.4	36.1
Red deer (Arman et al. 1974)	4	100	10.6 (8.5 to 13.1)[c]	7.8 (7.1 to 8.6)[c]	4.4 (4.4 to 4.5)[c]	23.9 (21.1 to 27.1)[c]
Fallow deer (Jenness and Sloan 1970)	1	1	12.6	6.5	6.1	25.3
White-tailed deer (Silver 1961)	1	4	10.4 (7.5 to 18.0)	10.3 (8.8 to 11.5)	3.0 (2.2 to 3.8)	25.2 (22.2 to 33.5)
Black-tailed deer (Kitts et al. 1956)	2	3	10.4 (10.2 to 10.5)	8.9 (8.1 to 9.6)	4.4 (3.9 to 4.7)	25.0 (24.5 to 25.4)
Domestic cow (Wright et al. 1939)	130	1,998	4.4	3.8	4.9	13.8
Domestic goat (Wright et al. 1939)	≥ 326	From 18 investigators	4.1	3.7	4.2	12.9
Domestic sheep (Wright et al. 1939)	2	Regular intervals	6.2	5.4	4.2	17.1

Note: References and ranges of variation are in parentheses. Ranges include whole lactations except for white-tailed deer and black-tailed deer (samples include early and end), and blackbuck antelope (unknown stages). Ben Shaul (1962) does not report numbers of animals or numbers of samples.

[a]Protein reported for only 2 blackbuck samples.

[b]Eland ranges are for monthly averages; protein range uncertain because 1 casein value is missing.

[c]Red deer ranges are for stage averages: 3 to 30 days, 31 to 100 days, > 100 days.

taxonomic relatedness (Ben Shaul 1962). Species like blackbuck (*Antilope cervicapra*), pronghorn (*Antilocapra americana*), and white-tailed deer (*Odocoileus virginianus*) that nurse their young only during brief, widely spaced care sessions also yield concentrated milk high in fat content (Ben Shaul 1962). This makes sense because the satiety value of fat exceeds that of any other milk component (Ben Shaul 1962). (For discussions of behavioral components of nursing and formula amounts that may be appropriate for a hand-raised fawn as it grows, see chapters 5 and 10.)

Table 7.4 lists lifetime production statistics for eastern dama gazelles as observed during a 12-year study by the author on Texas ranches and supplemented with data from Speeg et al. (2014) and the summary in the addra studbook compiled by Petric (2012). A striking point is how old a herd male may be before his first offspring arrives. More than one Texas owner has almost given up and replaced a herd male that seemed perfectly good in every other way because three years seemed so long to wait before any of his fawns arrived. By two years old, a male may look mature. He has the thicker body and neck of the typical grown male, but he may still be establishing his permanent dentition. He may be capable of breeding successfully at 17 to 30 months

Table 7.4. Lifetime production statistics for dama gazelles, using female reproductive life starting at 1.5 years old

REPRODUCTIVE ASPECT	QUANTITY
Life span expected on rangeland or in parks	7 to 12 yrs
Life span maximum reported in 2 parks	11 and 15 yrs
Life span maximum in zoo	19 yrs
Sexual maturation for female	13 to 17 mos, as low as 10 or 11 mos possible
Sexual maturation for male	17 mos
Female age at first birth	1.5 to 2.5 yrs
Male age at first birth	2 to 3 yrs
Gestation*	6 to 6.5 mos
Minimum interval expected between fawns	7 to 7.5 mos
Female breeding life expected to end*	14 yrs
Average fawns per year recorded for an 8-year-old rangeland female (fawning started at 2.5 years of age)	1.3
Possible number of fawns expected for:	
Female living to 7 years	7
Female living to 10 years	11
Female living to 12 years	13
Female living to 14 years	16
Female living to 19 years	16 to 22
(age in years – 1.5 years) x 1.3	

Source: Mungall data; Speeg et al. 2014; summary in Petric 2012.

*Reports summarized in literature (from Petric 2012 on gestation and breeding life, and from RZSS and IUCN Antelope Specialist Group 2014 on gestation).

Figure 7.20. This senior female, as evidenced by the long horns with long shanks as well as by the long, white face, has a newborn fawn at her feet (photo by Kathryn Kyle, courtesy of Kyle Wildlife, Texas, USA).

of age, but that still leaves the gestation period of 6 to 6.5 months (Petric 2012, RZSS and IUCN Antelope Specialist Group 2014) before there can be any births. Females, on the other hand, typically start their history of births at about 1.5 to 2.5 years of age. If gestation lasts 6 to 6.5 months, that means a female is sexually mature as young as 11.5 months old (while still an adolescent) to 24 months old (when just becoming a full adult). The 16- and 17-month minimum female ages at first parturition reported for two park-type properties (Speeg et al. 2014) indicate ages when first bred as low as 10 to 11 months, assuming a 6-month gestation. In species like the smaller Indian blackbuck antelope, females mature faster than males both sexually and physically (Mungall 1978). For sexual maturity this seems true for dama gazelles also, even though in dama gazelles like those on ranches progression through physical age categories based on body conformation and horn form track about the same for males and females. For both these species, males and females continue breeding successfully into old age (fig. 7.20).

Longevity

The long time before a dama gazelle male starts breeding mirrors the generally long life span of this species. It surprises some owners that 10- and 12-year-old dama gazelles out in Texas pastures are more the rule than the exception. One herd male that hung on to his territorial status, although retaining less land, until a lethal fight at 12 still had a reasonable set of teeth. At 9 years old, he had had a serious abscess with significant bone recession around his right lower molars, but his teeth still had effective, well-aligned wear surfaces when he died.

With close care and feeding in zoos, many species tend to live longer than expected either in pastures or in the wild. Table 7.4 shows that the dama gazelle, one of the tallest gazelles, has a particularly impressive zoo longevity record. The 2012 studbook for the eastern race of dama gazelles as kept by zoos in North America states that zoo dama gazelles living into their late teens are not exceptional, even though reproduction generally ends by the time the animals are 14 years old (Petric 2012). A zoo dama gazelle female that had recently given birth at 19 was euthanized because she was declining with old age. By contrast, zoo records for the smaller Indian blackbuck antelope indicate 16 years as the biological life span limit (Flower 1931, Crandall 1964, Mungall 1978).

Value of Benchmark Information

With these sorts of statistics, a wildlife manager can chart the progress of a dama gazelle population. Aspects that are going well can be clarified and problem areas identified. This helps ranch managers see what issues might need attention. Conservationists concerned about native populations can raise similar questions to see whether there are issues that can be addressed for the benefit of the population.

Chapter 8

Horns and Color Notes

Elizabeth Cary Mungall

A

B

Figure 8.1. Variation in the shape of the neck patch and horns can identify many individuals: *A*, especially among eastern dama gazelles (photo by Elizabeth Cary Mungall, courtesy of Kyle Wildlife, Texas); and *B*, somewhat among mhorr gazelles (photo by Gerardo Espeso, courtesy of Estación Experimental de Zonas Áridas, Spain).

Horns are often of general interest, but why write at length about coat color, too? Much of the discussion about dama gazelles has to do with color. It is involved in opinions about taxonomy, distribution, management, and identification. In a very practical way, color can be taken with horn shape as the "natural markings" used to identify individual animals in order to determine the dynamics of social behavior (fig. 8.1). The striking colors of dama gazelles fascinate people even apart from the efforts of researchers to determine whether various aspects of color have meaning for the animals themselves.

Coat color and horn shape are the two most conspicuous ways that dama gazelles differ from each other. While horn shape varies within the species in general, typical coat color shifts with geographic region. As already pointed out, these regional coat color variations have been used to define dama gazelle subspecies: *Nanger dama ruficollis* in the east, sometimes called addra, *N. d. dama* for the central dama gazelle, and *N. d. mhorr* for the mhorr gazelle in the west (Cano Perez 1984; also see chapter 2).

Horn Shape

All dama gazelles grow relatively short but well-formed and useful horns. Female horns are thinner than male horns. For size ranges, see chapter 7. Horns are S shaped as seen from the side (figs. 8.2*A* and 8.3*A*). From the front, classic male horns have gentle curves as in figure 8.2*B*. Tips point distinctly inward in young adults but ordinarily turn more forward in males as growth continues. Forward-facing points

A

B

Figure 8.2. Classic horn shape for fully grown adult male dama gazelles: *A*, pronounced S shape in side view; *B*, tips point forward in front view (photos by Elizabeth Cary Mungall, courtesy of Kyle Wildlife, Texas, USA).

A

B

Figure 8.3. Classic horn shape for adult female dama gazelles: *A*, shallow S shape in side view; *B*, tips point inward in front view (photos by Elizabeth Cary Mungall, courtesy of Kyle Wildlife, Texas, USA).

facilitate the characteristic dama gazelle raking attacks such as that against the haunches of a fleeing male, as shown in figure 6.5. The tips of female horns curve inward (fig. 8.3*B*).

As expected, some males and females have wider horn spreads than others, and there are further variations, although these are not as frequent (figs. 8.4 and 8.5). The horns of a very old female may even

A

B

C

D

E

Figure 8.4. Variations in male horn shape: *A*, wide spread; *B*, round and high above the head; *C*, angular; *D*, swept back from base; *E*, crooked (photos *A*, *D*, and *E* by Elizabeth Cary Mungall and *B* by Kathryn Kyle, courtesy of Kyle Wildlife, Texas, USA; photo *C* by Christian Mungall, courtesy of Safari Enterprises, Texas, USA).

Figure 8.5. Variations in female horn shape: *A*, wide spread; *B*, tips close together; *C*, parallel; *D*, crooked (photos by Elizabeth Cary Mungall, courtesy of Kyle Wildlife, Texas, USA).

suggest the beginning of an open spiral. Occasional males show a particularly tall and round horn shape (fig. 8.4*B*). Heredity may well account for an angular variant (fig. 8.4*C*) among males in a small population of mhorr gazelles gathered onto a single ranch in Texas. However, this angular shape is not confined to mhorr gazelles.

Inheritance may well play a role in defining horn shape in general. For example, a conspicuously wide-horned eastern dama gazelle female in Texas produced wide-horned male and female offspring, all sired by a classic male. A prominently crooked-horned eastern gazelle female (fig. 8.6*A*) in another herd produced an exaggeratedly crooked-horned male—not the result of injury (fig. 8.6*B*). Her other offspring with the same classic sire were normal. Females often have one horn angled slightly differently from the other, but males rarely do.

A very few males have both horns angled back so far from the base that they look as if they almost

A

B

Figure 8.6. Right horn distinctly lower than the left on both *A*, a mother and *B*, her son (photos by Elizabeth Cary Mungall, courtesy of Kyle Wildlife, Texas, USA).

curve around the ears (fig. 8.4*D*). Obviously, injuries can result in abnormal angles as well as in one-horned, or even hornless, individuals. One female with particularly thick horns grew a tiny third horn from one horn base (fig. 8.7*A*). It looked as if the base might have suffered a minor split, a mishap likely for other horned ungulates, too (see the dorcas gazelle, *Gazella dorcas*, in fig. 8.7*B*). This dama gazelle's thin third horn finally broke off. Females born into a herd for which supplemental feed such as protein pellets is routinely available seem to grow thicker horns.

A

B

Figure 8.7. Extra curl of horn: *A*, on an adult female dama gazelle (photo by Elizabeth Cary Mungall, courtesy of Kyle Wildlife, Texas); *B*, similar case for a dorcas gazelle male (photo by John Newby, copyright © John Newby/Sahara Conservation Fund, Chad).

Coat Color

The mhorr gazelle gets its name from the Arabic word *mohor* for the red-brown color of a chestnut colt of a domestic horse (Abdelkader Jebali, pers. comm.). This subspecies, *N. d. mhorr*, defines the basic dama gazelle pattern (figs. 8.4*C* and 8.8). Its rich chestnut coat—often tending toward dark red—has sharp divisions where this color meets white. Some individuals are of a lighter phase because the white reaches a little higher on the sides and the color on the haunches is somewhat restricted, but the general pattern is clear. Also, the haunch color still shows a diffuse indentation on its lower border that creates a double-lobed effect. This leaves a "ham" shape on part of the hind leg.

Map 3.1 locates the known remnants of the three races of today's dama gazelles (*mhorr*, *dama*, and *ruficollis*) in relation to the extent of their native range during the 1800s. (For further discussions of distribution, see chapters 1, 2, and 20.) Table 8.1 summarizes the changes in coat color from northwest to south and east.

The central *N. d. dama* is probably intermediate in variation because it grades into *N. d. mhorr* in the west and *N. d. ruficollis* in the east. In particular, the haunch color varies in the amount of indentation—from a little, as on the light-phase *mhorr*, to a greater intrusion of white that makes the haunch color into a haunch stripe (fig. 8.9). In either case, a slender projection from this haunch stripe to the hocks is a hallmark of *N. d. dama.* (For a diagram of body parts such as "hock" and color marks such as the haunch stripe, see the diagram in appendix 2.) Expected coat color is a bright chestnut.

Farther east, *N. d. ruficollis* shows the greatest color variation. Now, white hairs mix with chestnut hairs on the back and sides. This creates a roan effect, which sometimes looks pinkish (fig. 8.10). Orange chestnut on the neck grades on the back and upper sides into some amount of a variable shade of

Figure 8.8. Maximum dama gazelle color shown by a mhorr gazelle male (photo by Christian Mungall, courtesy of Safari Enterprises, Texas, USA).

Table 8.1. Shift in distinctive dama gazelle coat color characteristics across a selection of localities where the species is native

CHARACTERISTIC	PLACE (from west to east)						
	Western Sahara	**Tamesna**	**Aïr-Ténéré**	**Termit**	**Manga**	**OROA***	**Kordofan**
Red-brown color	X	X	X				
Orange-chestnut color				X	X	X	X
Small neck patch	X	?					
Enlarged neck patch		?	X	X	X	X	X
Triangular haunch color	X	?					
Broad haunch color, vertical to hock	X	X	X	X			
Thin haunch stripe, vertical to hock		X	X	X	?		
Thin haunch stripe, little or no vertical					X	X	
No haunch stripe						X	X
Side color division horizontal, sharp	X	X	X	X			
Side color division horizontal, diffuse					?	X	
Side color division diagonal					X	X	X
Side color same as neck	X	X	X	X	X		
Side color lighter than neck					X	X	X
Diffuse saddle patch					X	X	X
Virtually no color other than on neck						X	X

Note: Characteristics indicated by recent survey photographs supplemented at range extremes with animals in breeding centers (west) and historical reports (east).

*OROA stands for Ouadi Rimé–Ouadi Achim Game Reserve in Chad.

Figure 8.9. Variable haunch color shown by a group of central dama gazelles on the Termit Massif, Niger (photo by Thomas Rabeil, copyright © Thomas Rabeil/Sahara Conservation Fund).

Figure 8.10. Three bachelor age-mates, two with classic color and one exceptionally dark (photo by Elizabeth Cary Mungall, courtesy of Kyle Wildlife, Texas, USA).

A

B

Figure 8.11. Exceptionally light subadult male photographed twice on the same morning: *A*, looking white in the sunlight, but *B*, showing a saddle patch when in the shade (photos by Elizabeth Cary Mungall, courtesy of Kyle Wildlife, Texas, USA).

roan: anything from dark roan to so light that the glare of the sun on a glossy coat easily makes it look white (fig. 8.11).

Particularly light dama gazelles were reported as typical on the eastern edge of the dama gazelle's distribution in Kordofan, part of today's Sudan (see chapter 4). Intriguingly, a "white" dama gazelle does not necessarily have light parents. Although the first fawn of the wide-horned female mentioned above (which was mated with a classic *ruficollis* male) matured into one of the darker animals in her herd, looking almost exactly like her mother, subsequent progeny included one "white" male—the lightest individual ever seen in its ranch population—as well as both regular roan and light roan offspring (Mungall 2010) (fig. 8.12).

Yellowish color

In the same Texas *ruficollis* herd, with the same herd sire, a female whose coat took on a yellowish-tan

A

Figure 8.12. Example of extreme color variation within a family: *A*, sire; *B*, dam; *C*, a daughter very like her mother; *D*, an exceptionally light son (photos by Elizabeth Cary Mungall, courtesy of Kyle Wildlife, Texas, USA).

B

C

D

A

B

Figure 8.13. *A*, Mother that tended to have *B*, particularly dark sons (photos by Elizabeth Cary Mungall, courtesy of Kyle Wildlife, Texas, USA).

cast before fall, when the sun-bleached hair was shed and the coat was renewed, gave birth to the darkest male in the group (fig. 8.13). Like his mother, he had color extending in an oblong shape over the maximum *ruficollis* amount for back and sides. Unlike his mother, he lacked the diffuse, light "saddle patch" so common in *ruficollis. N. d. mhorr* lacks any saddle patch, and *N. d. dama* is said not to show it either (Cano Perez 1984). The only other hints of a yellowish coat in this population of approximately 85 *ruficollis* after changing from neonatal colors were in one female and one male. The female was never thrifty (snakebite was suspected), had a rough, dull coat, and died young after an injury. The other yellowish youngster, a male, was surprisingly stocky as a fawn and had only half a tail even from the time he was first seen as a neonate. (For a photograph of this male as a subadult, see chapter 2.) As mentioned in chapter 2 on taxonomy and distribution, a yellowish tint occurs very occasionally on museum skins from native habitat and appears to be related to a different crystalline structure that can develop in the pigmented hairs (Cano Perez 1984).

Haunch stripe

As well as extreme variation in the amount of body color on the back and sides, there is extreme variation relative to the development of the haunch stripe (Mungall 2010). Some individuals lose it entirely as they mature. This seems more common among males. Nevertheless, there are males with a clear haunch stripe, and some females also have white

Figure 8.14. Fading of *ruficollis* haunch stripe with age: this adult female had a strong, wide haunch stripe in 2005 but only a light mark—particularly the horizontal part—by mid-2008; this was especially so here in 2013, when she was old (photo by Elizabeth Cary Mungall, courtesy of Kyle Wildlife, Texas, USA).

Figure 8.15. Female eastern dama gazelle with exceptionally wide haunch stripe (photo by Elizabeth Cary Mungall, courtesy of Kyle Wildlife, Texas, USA).

Figure 8.16. Female eastern dama gazelle with a rare downward extension of the haunch stripe that faintly connects with color extending upward from the hocks (photo by Elizabeth Cary Mungall, courtesy of Kyle Wildlife, Texas, USA).

haunches from the time they reach full adulthood. As with the main part of the body, the haunch stripe is apt to gain more white hairs with the years. If the haunch stripe is narrow to begin with, it may eventually fade out altogether. Even a female of prime age that had one of the darker and wider haunch stripes in her group (fig. 8.12*B*) lost most of this color mark with old age, as seen eight years later (fig. 8.14).

Exceptionally wide haunch stripes in the eastern dama gazelle (fig. 8.15) are likely to be short and blunt, without much of the short stretch of downward turn toward the end that the haunch stripe can have even among some *ruficollis* individuals.

Very occasionally (only 1 among approximately 85 *ruficollis* in a population that had been observed for more than 10 years), this downward projection is a slender, light tan extension as far as the hocks (fig. 8.16). Thus, this very exceptional case looks like a faint reminder of the tail of color to the hocks that is a distinguishing feature of *N. d. dama.*

Nanger rump patch and haunch stripe

When there is a long stripe on the haunches, this color mark appears to develop from a merger of two different elements (fig. 8.17). There is the dark rim of the white gazelline rump patch, a feature often

A

B

Figure 8.17. Adolescent eastern dama gazelles showing the two elements that can form the haunch stripe when it has a downward projection: *A*, young adolescent with light tan still covering much of the haunch; *B*, adolescent (stotting) almost fully changed to adult coloration except for tan edge of gazelline rump patch (photos by Elizabeth Cary Mungall, courtesy of Kyle Wildlife, Texas, USA).

accentuated in gazelles. In Grant's gazelle (*Nanger granti*), this edge is even likely to go black. Among eastern, and especially central, dama gazelles, the dark vertical rim makes the downward projection of the dama gazelle haunch stripe. Then there is the dark of the upper margin of color on the hind legs as the distinctive forward prongs of the *Nanger* rump patch fade in. This chiefly horizontal element becomes the haunch stripe common in eastern dama gazelles (fig. 8.17*B*).

As long as it lasts, the haunch stripe creates a vague outline of the *Nanger* rump patch even as the rest of the haunch color fades away. For central and eastern dama gazelles, this is not so much a darkening of these edges as it is a retention of the two dark borders while the main part of the hind legs lightens. Finally, the main part fades out altogether and the border remnants seem to grow together on the individuals in which they are retained. On the western dama gazelle, the classic mhorr gazelle, the *Nanger* patch fades in and joins with the white gazelline portion of the rump patch, but the main area of color on the hind legs never fades. There is no extra darkening on the borders of the rump patch. Soemmerring's gazelle (*Nanger soemmerringii*), like mhorr gazelle, has a large *Nanger* rump patch without extra darkening on the edges. Grant's gazelle has black on the vertical edge of the rump patch, giving the body color heightened contrast with the rump patch. The white space of the rump patch stays relatively restricted in Grant's gazelle. Figure 8.18 gives a comparison of adult coloration for all three *Nanger* species.

Figure 8.18. Comparison of adults of all three *Nanger* gazelle species: *A*, mhorr gazelle male (*left*) and female; *B*, eastern dama gazelle male behind female; *C*, Soemmerring's gazelle males; *D*, Grant's gazelle male (grazing) and female (photos by Elizabeth Cary Mungall, *A* and *C* courtesy of Safari Enterprises, Texas, USA; *B* courtesy of Kyle Wildlife, Texas, USA; *D* courtesy of Tulsa Zoo, Oklahoma, USA).

Figure 8.19. Small neck patch characteristic of fawns of all dama gazelle subspecies (photo by Cole L. Reid, courtesy of Stewards of Wildlife Conservation, Texas, USA).

Figure 8.20. Common large neck patch with very few dark hairs above (*back left*, female) and the rare separated neck patch (*right front*, female) (photo by Christian Mungall, courtesy of Kyle Wildlife, Texas, USA).

Figure 8.21. Mhorr gazelle female of the dark phase but without a neck patch (photo by Abdelkader Jebali, courtesy of Guembeul Fauna Reserve, Senegal).

Neck patch

As with other color differences, variations in the neck patch and associated areas are greatest in *N. d. ruficollis* (fig. 8.1*A*, *B*). Although all eastern, western, and presumably central dama gazelle fawns are routinely born with a small neck patch (fig. 8.19) similar to that characteristic of full-grown mhorr gazelles (fig. 8.4*C*), this patch ordinarily expands in *N. d. dama*, and even more so in *N. d. ruficollis*, as the animals gain their adult colors (Cano Perez 1984, Mungall, pers. observation).

One conspicuous neck patch variation sometimes recurs in succeeding generations. A daughter of a male with a neck that was almost white from the neck patch upward was similar in spite of her dam having a clearly defined neck patch (fig. 8.20 *left*). Whether it be in *N. d. mhorr*, *N. d. ruficollis*, or, presumably, *N. d. dama*, the most striking neck patch oddity is the occasional case of displacement, or even total separation, near the center (fig. 8.20 *right*). To date, these displacements or separations have not been noted to continue down lineages. Only once has a dama gazelle been found without any neck patch at all (observation by Abdelkader Jebali at a North African reintroduction center for mhorr gazelle) (fig. 8.21).

Age effects and significance of color marks

Ungulate color elements that emphasize one body part or another are expected to have significance. Do the bands down the fronts of the legs, with their roan or chestnut expansions on the fetlock joints, make it easier, in spite of the glare from the white on the rest of the legs, for another dama gazelle to see that a conspecific is approaching? Do the light saddle patch of *ruficollis* and the progressively lighter average body color to the east across the distributional area of the species indicate an increased need to moderate heating from solar radiation?

Certainly, dark hair on the foreheads of immatures and younger adults does make the horns of both males and females look longer (fig. 8.22). The faces of older adult *ruficollis*, *mhorr*, and probably *dama* go white (fig. 8.23). By then, horns have reached their full growth. And what about the neck patch?

Figure 8.22. Dark hair on the foreheads of younger dama gazelles, like this male, makes horns look longer (photo by Elizabeth Cary Mungall, courtesy of Kyle Wildlife, Texas, USA).

Figure 8.23. Progressive whitening of the face: *A*, three generations of the eastern dama gazelle showing white face of (*left to right*) the wild-caught matriarch compared to a young adult and an adolescent (photo by Elizabeth Cary Mungall, courtesy of the San Antonio Zoo, Texas, USA); *B*, lightening of the face with age in mhorr gazelle as shown by (*right to left*) an adult female more than 7 years old, a fully adolescent male 5 to 6 months old, and a newly adolescent female of 3 months old (photo by T. Abáigar/CSIC, courtesy of Estación Experimental de Zonas Áridas, Spain).

The unique differences among neck patch shapes that develop as adult colors come in remain throughout life. Positioned on the chestnut neck rather than on the roan body, even the neck patch of *ruficollis* retains its special characteristics in spite of the way the roan may lighten, the outlines of *ruficollis* color on the back and sides may become less distinct over the years, and the haunch stripes of *ruficollis* may fade.

The white neck patch is probably the single most distinctive color marking of the dama gazelle. It is displayed by a suitor to any female he may be following as he elevates his head. Similarly, any male or female using the head-flagging dominance display shows the neck patch when the nose is turned toward a target individual. In this regard, it is worth remembering that, unlike the females of some gazelles, female as well as male dama gazelles have fully developed horns. As shown in the chapter on growing up and growing old (chapter 7), a female dama gazelle will sometimes engage horns with another dama gazelle—male or female.

In interspecies situations, dama gazelles often associate with whatever other gazelles or similar animals are in their area. In native habitat, this is frequently the dorcas gazelle (*Gazella dorcas*). As exotics in Texas, this is likely to be Thomson's gazelle (*Gazella thomsoni*) or blackbuck antelope (*Antilope cervicapra*) (fig. 8.24).

Figure 8.24. Dama gazelles often join other gazelles or gazelle-like species: *A*, dama gazelle males with a group of Thomson's gazelles (photo by Elizabeth Cary Mungall, courtesy of Stewards of Wildlife Conservation, Texas, USA); *B*, one of several dama gazelle females and young traveling with a group of blackbuck antelope (photo by Elizabeth Cary Mungall, courtesy of Stevens Forest Ranch, Texas, USA).

Figure 8.25. Herd of exotic addax (photo by Elizabeth Cary Mungall, courtesy of Stevens Forest Ranch, Texas, USA).

Figure 8.26. Herd of exotic scimitar-horned oryx (photo by Elizabeth Cary Mungall, courtesy of Stewards of Wildlife Conservation, Texas, USA).

Which animals have a neck patch and which do not may be significant for species recognition. This may be even more important for young fawns, since they all look very much alike except for the neck patch on dama gazelles (fig. 8.19). Larger adults most likely to be in the same area near dama gazelles are the scimitar-horned oryx (*Oryx dammah*) and the addax (*Addax nasomaculatus*). Addax use the desert more, including sandy expanses, and look only vaguely like dama gazelles (fig. 8.25).

Scimitar-horned oryx, on the other hand, are more intermediate in habitat preference and often frequent localities being used by any of the dama gazelle subspecies even though the two species do not team up together the way dama gazelles and other gazelle types do. With their chestnut neck (no neck patch) and white-to-roan backs and sides, scimitar-horned oryx are colored very like *N. d. ruficollis* (fig. 8.26). Put them together in a zoo desert exhibit, and keepers cringe at the inevitable exclamations of "Look at those mothers and their babies!" Real scimitar-horned oryx calves are an orange tan with white above the hoofs—vaguely like dama gazelle fawns in color, but with no neck patch, a larger size, and stockier body contours.

Individual identification

Regardless of the significance dama gazelles may place on the neck patch they carry, researchers find its unique variations a most valuable tool. These neck patch differences can often serve to identify individual animals without the risk of catching them in order to place artificial tags. Although it can be difficult to keep track of juvenile dama gazelles as they separate increasingly from their mothers and go through their transition to adult coloration, as soon as their "final" colors become established, there is a good chance of finding peculiarities of the neck patch that will allow the animal to be identified. (For a discussion of color changes with maturation, see chapter 7.) When neck patch peculiarities are pronounced, the animal may be identifiable for the rest of its life. Otherwise, these neck patterns may be enough to allow identification during the gap between adolescence when the adult colors become apparent and adulthood when the full horn shape can aid in individual identification. The combination of neck patch shape and horn shape allows many individual dama gazelles to be recognized when they would otherwise be just one animal among many that share similar neck patch or horn shape characteristics.

Chapter 9

Food Habits of the Dama Gazelle

Abdelkader Jebali and Elizabeth Cary Mungall

The rapid decline of dama gazelles in the wild and the relatively late realization of this problem has prevented a thorough investigation of the biology, ecology, and ethology of this species in the wild. Existing data are rare, old, and scattered and often deal with hunting or distribution. However, several studies on physiology, behavior, home range patterns, and genetics have been undertaken on captive populations. These projects have focused mainly on mhorr gazelle (*Nanger dama mhorr*), the subspecies native to Western Sahara (Cano Perez 1990, Cassinello and Pieters 2000, Cassinello 2005, Arroyo Nombela et al. 1990, Pickard et al. 2001), but both wild-caught and zoo-born, as well as pasture-raised, members of the eastern subspecies, also known as addra (*Nanger dama ruficollis*), all originating from Chad, have also been studied (Mungall 1980, 2007b, 2013, 2015; Mungall et al. 2007). The dama gazelle's diet is one of the subjects that has suffered from this lack of attention. Exceptions are often circulated in reports that are not always accessible to everyone (Newby 1974, Grettenberger and Newby 1986, Dragesco-Joffé 1993, Poilecot 1996b).

The reintroduction and repatriation of the mhorr gazelle in some countries of its former range have allowed collection of a little more data on food preferences (Cammaerts 2003, Jebali 2003) (fig. 9.1). Release of the eastern dama gazelle in exotic habitats such as ranches in Texas, in the United States (figs. 9.2 and 9.3), has shown the food adaptability of the species to different forage plants. The data presented here are a compilation of observations undertaken both in North Africa and in exotic environments.

Figure 9.1. Mhorr gazelle male at a reintroduction center in North Africa (photo by Abdelkader Jebali, courtesy of North Ferlo Fauna Reserve, Senegal).

Figure 9.2. Prime habitat for exotic dama gazelles on the Edwards Plateau of Central Texas (photo by Elizabeth Cary Mungall, courtesy of Kyle Wildlife, Texas, USA).

Figure 9.3. Preferred site in a West Texas pasture (photo by Elizabeth Cary Mungall, courtesy of Stevens Forest Ranch, Texas, USA).

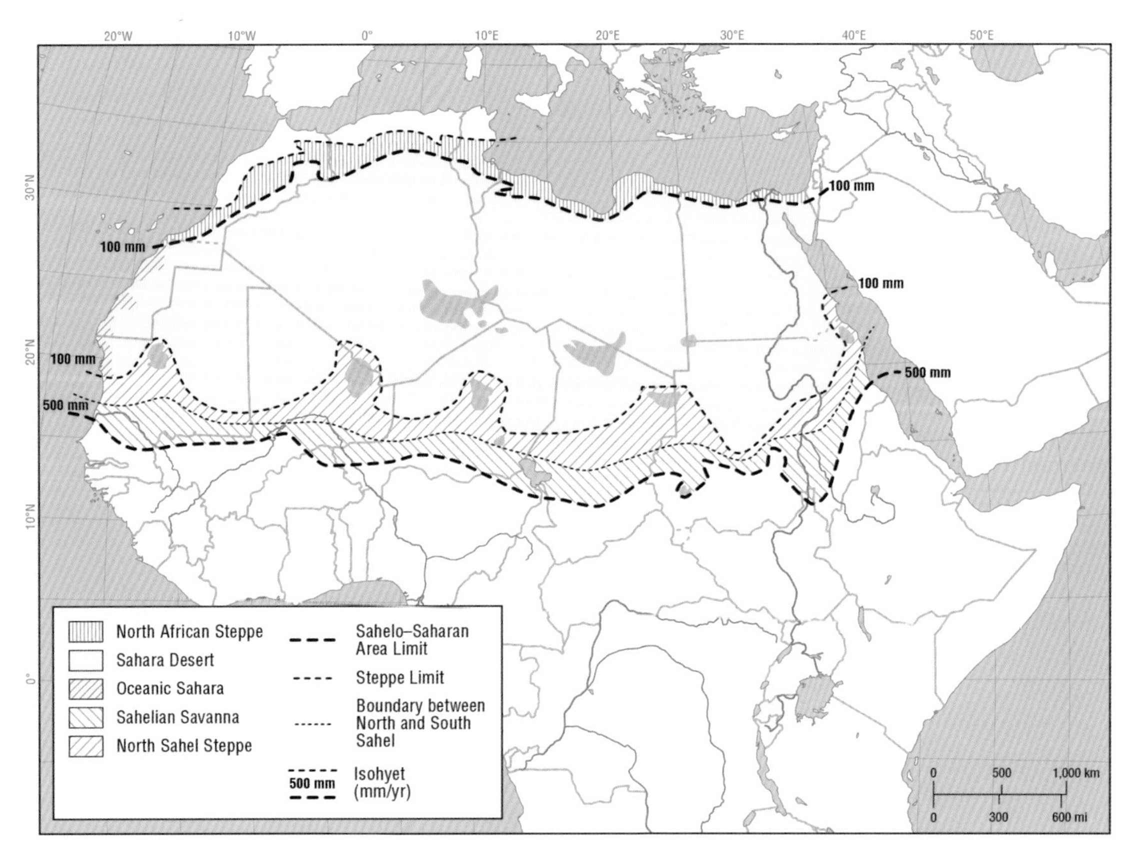

Map 9.1. Arid zones in North Africa showing the extent of the northern and southern Steppe and Sahel belts in relation to the Oceanic Sahara and Sahara Desert (by Abdelkader Jebali, 2007).

Figure 9.4. Typical appearance of what was prime dama gazelle habitat in Chad, showing dry grass plains scattered with trees and bushes (photo by John Newby, copyright © John Newby/Sahara Conservation Fund, Chad).

Dama Gazelle Food Habits in Natural African Habitat

The principal habitat for dama gazelles is the steppe and savanna borders of the Sahara. A belt of this semiarid grassland runs both north and south of the desert (map 9.1). The southern Sahelian Zone is substantially wider than the steppe belt north of the Sahara. To the south, the expanses of grass grade into savanna and then woodland.

Figure 9.4 shows the grassy plains studded with occasional trees and bushes typical of what used to be the best dama gazelle habitat before domestic livestock largely pushed wild dama gazelles into marginal areas. In the heat of the dry season, dama gazelles still seek shade and browse in the brush along seasonal watercourses. Rains that occasionally fall in the desert lure dama gazelles into the Sahara for the brief flush of new growth.

In the Sahelian Zone, there is a rainy season that lasts roughly three months and a dry season that spreads over nine months, and the dama gazelle appears to be adapted in its diet according to changes in these two seasons. Grettenberger and Newby (1986) identified at least 14 species of plants palatable for this gazelle in the Aïr and Ténéré National Nature Reserve, Niger, between March 1983 and March 1984 (table 9.1). During that period of drought, the gazelles showed a clear preference for *Acacia tortilis* (taken as a synonym of *Acacia raddiana*), *Balanites aegyptiaca*, and *Chrozophora brocchiana*. In a second study, they consumed *Maerua crassifolia* and *Leptadenia pyrotechnica* (Poilecot 1996b). Thus, this gazelle was behaving as a browser, taking most of its food from the leaves of trees and shrubs. This observation has been confirmed by several authors (Newby 1974, Poilecot 1996b, Dragesco-Joffé 1993, Jebali 2003, 2008).

In Ouadi Rimé–Ouadi Achim Game Reserve, Chad, Newby (1974) considered the dama gazelle to be more of a browser than the dorcas gazelle (*Gazella*

Table 9.1. Plant species dama gazelle (*Nanger dama*) preferred as food in Aïr and Ténéré National Nature Reserve, Niger, March 1983–March 1984

Species	Number of plants eaten/number of plants observed*
Leptadenia pyrotechnica	8/8
Maerua crassifolia	4/4
Balanites aegyptiaca	20/23
Acacia tortilis (*A. raddiana*)	31/41
Chrozophora brocchiana	13/42
Aerva javanica	2/65
Panicum turgidum	2/109
Calotropis procera	1/1
Citrullus colocynthis	1/1
Pergularia tomentosa	1/1
Acacia albida	0/1
Fagonia bruguieri	0/1
Farsetia ramosissima	0/1
Stipagrostis uniplumis	0/1

Sources: After Grettenberger and Newby 1986, Poilecot 1996b, modified by A. Jebali.

*As an example, for *Leptadenia pyrotechnica*, 8/8 means that of 8 plants observed, the gazelle has touched 8; for *Acacia tortilis*, 31/41 means that of 41 specimens observed, the gazelle has touched only 31, and so on.

dorcas), and more of a grazer than the red-fronted gazelle (*Gazella rufifrons*). Newby reported 28 species of palatable plants eaten by dama gazelles, including a significant number of species of both grasses and forbs (forbs being nonwoody, often weedy, broad-leaved plants) (table 9.2).

In Senegal, where the western *mhorr* race was reintroduced in 1984 (Cano et al. 1993, Jebali 2003, 2008), the gazelle seemed well acclimated to the succession of seasons. It preferred wooded areas, especially in the dry season, when it was observed browsing the leaves and fruits of *Balanites aegyptiaca* and especially of *Acacia raddiana*, *Acacia seyal*, and *Leptadenia hastata* (fig. 9.5). During the rainy season, the mhorr gazelles gradually left wooded areas and settled in places that were mostly open and covered by herbaceous plants (Jebali 2008). During the 2005 rainy season, the mhorr gazelles were observed for almost three months, from July to September, on the "Plateau Nord," an open area covered with patches of grass (fig. 9.6).

When the rainy season was slow in coming or during drought, the mhorr gazelles used species not touched by livestock, such as *Calotropis procera*

Table 9.2. Dama gazelle (*Nanger dama*) diet in the Ouadi Rimé–Ouadi Achim Game Reserve, Chad

Browse species (13) (trees and bushes)	Grazed species (15) (grasses and forbs)
Acacia senegal	*Aristida mutabilis*
Acacia tortilis (A. raddiana)	*Aristida pallida*
Balanites aegyptiaca	*Blepharis linariifolia*
Boscia senegalensis	*Boerhavia repens*
Cadaba farinosa	*Borreria radiata*
Capparis corymbosa	*Citrullus colocynthis*
Capparis decidua	*Commelina forskalaei*
Chrozophora senegalensis	*Cucumis melo*
Grewia tenax	*Indigofera aspera*
Grewia villosa	*Limeum viscosum*
Guiera senegalensis	*Monsonia senegalensis*
Leptadenia pyrotechnica	*Panicum turgidum*
Maerua crassifolia	*Schmidtia pappophoroides*
	Tephrosia lupinifolia
	Tephrosia obcordata

Source: After Newby 1974, compilation by A. Jebali.

Figure 9.5. A splendid male mhorr gazelle browsing the leaves of *Leptadenia hastata* (photo copyright © Abdelkader Jebali, courtesy of North Ferlo Fauna Reserve, Senegal).

Figure 9.6. An open area with patches of grass, the "Plateau Nord" is favored by mhorr gazelles in the rainy season (photo copyright © Abdelkader Jebali, courtesy of North Ferlo Fauna Reserve, Senegal).

and *Euphorbia balsamifera.* The large dama gazelles browsed the leaves, flowers, and green fruits of *Calotropis procera* (fig. 9.7; table 9.3). They often visited *Euphorbia balsamifera* to tear the bark and access the sap (Jebali 2008, Mbouyou Boulende 2011) (fig. 9.8).

Thanks to their long legs and neck, dama gazelles can also exploit the higher stages of plants by rising on their hind legs like goats, except that dama gazelles let their forelegs hang stiffly, rarely propping them on any possible supports (fig. 9.9). Rearing up to grab browse allows dama gazelles to reach leaves and pods that are not accessible to other species that share the same habitat.

When dama gazelles eat from tall vegetation such as bushes, as opposed to plucking forbs or flower

Figure 9.7. Male mhorr gazelle reaches for leaves and flowers of *Calotropis procera*, a plant used during dry times (photo copyright © Abdelkader Jebali, courtesy of North Ferlo Fauna Reserve, Senegal).

Table 9.3. Mhorr gazelle (*Nanger dama mhorr*) diet in Guembeul and North Ferlo Fauna Reserve, Senegal (reintroduced population)

Species (trees and bushes)	Palatable parts
Balanites aegyptiaca	Leaves, fruits
Leptadenia hastata	Leaves, buds
Calotropis procera	Leaves, flowers, fruits
Acacia seyal	Leaves, buds, flowers, and pods
Acacia raddiana (*A. tortilis*)	Leaves, pods
Prosopis chilensis	Only pods
Boscia senegalensis	Leaves
Commiphora africana	Leaves
Grewia tenax	Leaves
Guiera senegalensis	Leaves
Ziziphus mauritiana	Leaves

Source: After Jebali 2003, 2008.

A

B

Figure 9.8. *Euphorbia balsamifera* as used during dry times: *A*, bush; *B*, close-up showing patches of bark gnawed off branches (photos copyright © Abdelkader Jebali, Senegal).

Figure 9.9. Mhorr gazelle on its hind legs to access higher browse (photo copyright © Abdelkader Jebali, courtesy of Guembeul Fauna Reserve, Senegal).

Figure 9.10. Mhorr gazelles do much of their browsing at or above head height; here they are feeding on *Acacia seyal* (photo copyright © Abdelkader Jebali, courtesy of North Ferlo Fauna Reserve, Senegal).

Figure 9.11. Mhorr gazelles around a bush showing typical browsing arrangement for dama gazelles (photo by T. Abáigar/ CSIC, courtesy of Guembeul Fauna Reserve, Senegal).

heads from among grasses, they do much of their browsing at or above head height (fig. 9.10). When a group of gazelles forages among bushes, it is characteristic for two to four animals to space themselves around the same bush and all eat at once (figs. 9.10 and 9.11).

In Tunisia, where the same *mhorr* subspecies was released in Bou Hedma National Park (Kacem et al. 1994a; Abáigar et al. 1997, Wiesner and Müller 1998, Jebali and Zahzah 2013), Cammaerts (2003) identified 18 plant species palatable for the mhorr gazelles during April and May (table 9.4). The gazelles seemed to favor the park's *Acacia raddiana* (*A. tortilis*). However, they adapted their diet according to the amount of rainfall. Browsers before the rains, the mhorr gazelles changed their diet after some time. They almost became grazers, taking advantage of the growth of grasses rich in water and low in tannins (Cammaerts 2003).

Table 9.4. Mhorr gazelle (Nanger dama mhorr) diet in Bou Hedma National Park, Tunisia (reintroduced population)

Species (18)
Tree layer
Acacia raddiana (A. tortilis)
Shrub layer
Lycium arabicum
Periploca laevigata
Retama raetam
Rhus tripartita
Ziziphus lotus
Herbaceous layer
Anabasis oropediorum
Aristida obstusa
Aristida plumosa
Artemisia campestris
Avena alba
Cenchrus ciliaris
Cynodon dactylon
Diplotaxis harra
Echium hirtum
Eruca vesicaria
Stipa retorta
Tetrapogon villosus

Source: After Cammaerts 2003.

It is noteworthy that acacias are present in the dama gazelle diet both north and south of the Sahara. It is, moreover, in areas forested with *Acacia raddiana* that Valverde (1957) assigns the distribution of the mhorr gazelle in the Atlantic Sahara (in southern Morocco and northern Mauritania).

The dama gazelle seems well able to adjust its diet according to conditions. This must help during seasonal movements over large distances. Also, the species tends to adopt a mixed diet. It grazes more during the rainy season when the herbaceous layer is well furnished with grasses and forbs and changes to browsing more—sometimes exclusively—during the dry season when only shrubs and a few trees keep green foliage. This ability to adapt to different situations helps the dama gazelle acclimate to the conditions of captivity and to introduction into nonnative environments.

Dama Gazelle Food Habits in Nonnative Environments

Figure 9.2 shows a dama gazelle male in prime habitat in the Hill Country of Central Texas on the Edwards Plateau. Dama gazelles are confirmed browsers when kept as exotics on Texas rangeland. They will take a few bites of grass, but not many. If exotic eastern dama gazelles, the *ruficollis* subspecies called addra, are left in a limited pasture with a grassy opening and brush around the edges, they create paths as they amble back and forth through the grass, but they leave most of it standing. Their attention centers on certain forbs as the seasons bring different species to attractive stages, and the gazelles crane their necks to get every last reachable leaf from trees and bushes. When they need even more height, they, like the mhorr gazelle in figure 9.9, rear up and may tread for a few moments to gain extra time to catch overhead leaves. This upright stance is one of the trademarks of the dama gazelle as well as one of the strategies that add flexibility to its food habits. Texas dama gazelles also show the same pattern seen at North African reintroduction sites of several animals surrounding a bush and all eating at once (figs. 9.10 and 9.11).

Table 9.5. Plants used conspicuously by exotic dama gazelles (*Nanger dama ruficollis*) on the Edwards Plateau in Central Texas, listed alphabetically within growth category

Species and groups of species (17)
Tree layer
Celtis laevigata (sugar hackberry)
Juniperus ashei (Ashe juniper)
Prosopis glandulosa (honey mesquite)
Prunus serotina (black cherry)
Quercus spp. (oaks)
Shrub and vine layer
Acacia rigidula (*Vachellia rigidula*) (blackbrush)
Berberis trifoliolata (agarita)
Smilax sp. (greenbrier)
Vitis candicans (mustang grape)
Forb layer
Cirsium spp. (thistle)
Cooperia drummondii (rainlily)
Euphorbia marginata (snow-on-the-mountain)
Ratibida columnaris (upright prairie-coneflower)
Solanum elaeagnifolium (silverleaf nightshade)
Verbesina virginica (frostweed)
Cactus layer
Opuntia spp. (pricklypear)
Epiphyte layer
Tillandsia recurvata (ball moss)

Source: Observations by Mungall.

Figure 9.12. Heavily used blackbrush (*Acacia rigidula*): *A*, bush in flower; *B*, close-up of flowers amid thorns (photos by Elizabeth Cary Mungall, courtesy of Stewards of Wildlife Conservation, Texas, USA).

Table 9.5 lists some of the plants of particular note used by dama gazelles in the Texas Hill Country. Like the western mhorr gazelles, dama gazelles of the eastern subspecies exotic in Texas make heavy use of an acacia (in this case blackbrush, *Acacia rigidula*, recently renamed *Vachellia rigidula*, fig. 9.12) where available and also the acacia-like mesquite (*Prosopis glandulosa*). Thus, certain kinds of plants as well as certain growth forms of plants are similar for both native dama gazelles and exotics.

Another behavior that expands the range of foods for dama gazelles is their habit of gnawing on bark and woody limbs. If left in limited pastures, dama gazelles will eventually finish off a favored bush, such as an agarita (*Berberis trifoliolata*), by repeated pauses to chew on its limbs. This continues long after all the leaves have been stripped. Eventually, little more than stumps or bare branches may remain (fig. 9.13).

And this is in a pasture where the animals are receiving supplemental feed during times when natural forage is in short supply. In one instance, a bottle-raised dama gazelle nearing three months of age was already stripping bark from a dry juniper post in her enclosure (fig. 9.14).

A

B

Figure 9.13. Agarita bush (*Berberis trifoliolata*): *A*, branch with ripening berries; *B*, bush gnawed and denuded by dama gazelles (photos by Elizabeth Cary Mungall, courtesy of Kyle Wildlife, Texas, USA).

Figure 9.14. Older dama gazelle fawn pulling bark off a dry juniper (*Juniperus ashei*) post in her enclosure (photo by Elizabeth Cary Mungall, courtesy of Kyle Wildlife, Texas, USA).

Figure 9.15. It would take a tall animal to reach this highly preferred greenbrier vine (*Smilax* sp.) growing atop a spiny agarita bush (photo by Elizabeth Cary Mungall, courtesy of Kyle Wildlife, Texas, USA).

A

B

Figure 9.16. *A*, Thistles (*Cirsium* spp.) are favorite food items; *B*, thistles surrounded by less-favored upright prairie-coneflower plants (*Ratibida columnaris*) (photos by Elizabeth Cary Mungall, courtesy of Kyle Wildlife Texas, USA).

When this gazelle was introduced to a selection of local forage species at four to five months old, she showed definite preferences for certain kinds of browse. Greenbrier vines (fresh or drying leaves and stems, *Smilax* sp.) was the definite favorite (fig. 9.15). Vines of mustang grape (but not the unripe grapes, *Vitis candicans*) were a close second. Next came sugar hackberry (*Celtis laevigata*) and then black cherry (*Prunus serotina*).

An agricultural agent in Texas once said that he wished he had a herd of dama gazelles that he could loan out to ranchers for weed control. Listening to the list of plants that dama gazelles have been seen eating regularly, he shook his head. The tough, spiny selections failed to match his impression of the seemingly soft little muzzles on the doe-eyed creatures he had driven past on his way through the ranch. He was not alone. When the operators of this exotics ranch decided to let their expanding nucleus of dama gazelles have more room, they spent tedious hours grubbing out thistles before opening the gates. So, what was the first thing the dama gazelles did? They scoured their new enclosure eating every last thistle they could find right down to the ground. Their mouths may be small and their noses velvety, but fibrous, spiny thistles (*Cirsium* spp.) are among their favorite foods (fig. 9.16).

Besides thistles and the holly-like agarita already mentioned, pricklypear cactus (*Opuntia* spp.) is another spiny dama gazelle food, although it is not eaten quite as often. Pads that have fallen to the ground—or been knocked off—are sometimes picked up and chewed (fig. 9.17). One dominant buck even threatened the others in his group and kept them away from the pricklypear patch he was eating from until he had had his fill.

Some plants have mainly their flowers taken. The stringy, invasive upright prairie-coneflower (*Ratibida columnaris*) still commonly grows in dama gazelle pastures, but it loses many of its tops (fig. 9.16*B*). Dama gazelles also select the blooms of silverleaf nightshade (*Solanum elaeagnifolium*), a member of a genus with species that are often toxic to other animals. Snow-on-the-mountain (*Euphorbia marginata*)

Figure 9.17. Pricklypear pads (*Opuntia* spp.) are sometimes chewed by dama gazelles (photo by Elizabeth Cary Mungall, courtesy of Kyle Wildlife, Texas, USA).

Figure 9.18. Snow-on-the-mountain (*Euphorbia marginata*): *A*, a mature plant; *B*, a buck coated with sap and dust after eating the plants (photos by Elizabeth Cary Mungall, courtesy of Kyle Wildlife, Texas, USA).

Figure 9.19. Frostweed (*Verbesina virginica*), also known as iceplant, growing thickly below an oak tree (photo by Elizabeth Cary Mungall, courtesy of Kyle Wildlife, Texas, USA).

is also a food in a genus noted for toxicity. Dama gazelles push in and browse from these plants when they grow tall and flower. During the summer season, the gazelles will emerge from stands of snow-on-the-mountain so streaked with the thick sap that dust sticks to their coats (fig. 9.18).

Interestingly, another member of this plant genus, *Euphorbia balsamifera*, has been noted to be eaten by dama gazelles during dry times in Senegal. Other plants are presumed to provide moisture. In Central Texas, patches of frostweed (*Verbesina virginica*) grow in the shade of many oaks. Fawns find cover there and their elders find food (fig. 9.19).

The epiphyte known as ball moss (*Tillandsia recurvata*) that falls from trees during wind storms is often searched out from the ground (fig. 9.20). Dama gazelles will also rear up and pull it from the branches.

Windfall can be a problem when a shower of coveted oak acorns (*Quercus* spp.) suddenly makes

Figure 9.20. Ball moss (*Tillandsia recurvata*) forms dense clusters where bare tree branches offer favorable sites (photo by Elizabeth Cary Mungall, courtesy of Kyle Wildlife, Texas, USA).

Figure 9.21. This juniper bush (*Juniperus ashei*) has been denuded by dama gazelles as far as they can reach, so now they return frequently to gnaw on the bare branches (photo by Elizabeth Cary Mungall, courtesy of Kyle Wildlife, Texas, USA).

this favored food available in quantity. A dominant male may monopolize it for himself. Dama gazelles found dead after gorging themselves have almost nothing but acorns in their rumen. Setting out piles of other attractive food when winds are heavy during years with abundant acorn crops is usually the best management response.

Browse from common trees and bushes includes not only oaks (*Quercus* spp.), hackberry (*Celtis* spp.), black cherry (*Prunus serotina*), and honey mesquite (*Prosopis glandulosa*), but also tough juniper growth (*Juniperus ashei*). If the lower branches of a juniper bush have been denuded of green, the bare branches are periodically chewed and thrashed (fig. 9.21).

But not all dama gazelle foods are tough or toxic. The soft flowers of eveningstar rainlily (*Cooperia drummondii*) are a great delicacy (fig. 9.22). When these flowers spring up after a rain shower, dama gazelles often eat every one they can find.

So, whether a staple or a special addition to their diet, dama gazelles can eat from ground level to well above their normal head height in order to make a meal of whatever browse and forbs are on offer. Both their readiness to take tough and spiny plants and their lower interest in grasses, except when at their freshest, probably reflect what might be available in a dry environment such as their African homelands. As also pointed out for the mhorr gazelle, this flexibility has been a further advantage in the dama gazelle's adaptation to life as an exotic in other countries. Similarly, it bodes well for dama gazelles being considered for reintroduction projects.

Figure 9.22. The flowers of eveningstar rainlily (*Cooperia drummondii*) are an "ice cream plant" for dama gazelles, greatly relished but never common enough to be a staple in their diet (photo by Elizabeth Cary Mungall, courtesy of Kyle Wildlife, Texas, USA).

Chapter 10

Hand Raising a Dama Gazelle

Elizabeth Cary Mungall

An examination of hand rearing for the critically endangered dama gazelle (*Nanger dama*) was initiated when a newborn eastern dama gazelle fawn (*N. d. ruficollis*) had to be withdrawn from its breeding group at Kyle Wildlife in Central Texas. Because many owners of exotic ungulates on Texas rangeland keep this species, guidelines for saving the occasional orphan can be particularly important for conservation. Notable benchmarks to do with the fawn's nutrition, locomotion, and socialization can signal progress. Similarly, aspects of particular significance for the handling of dama gazelles being hand raised have been identified.

The way Kathryn Kyle and Scott A. Smith kept track of progress for a dama gazelle fawn that had to be hand raised at their ranch has furnished the majority of the following information. Additional experiences were contributed by Texas exotics breeder Douglas E. Smith of Bear Creek Exotics, who has hand raised several eastern dama gazelle fawns.

Feeding

The small female study fawn was withdrawn on March 2, 2015, for hand raising because the weather was turning cold and the mother had not cleaned her newborn even after 3 hours. Thus, the fawn was still wet. The fawn's starting weight was estimated at 2 to 3 kg (5 to 6 lb), about half the normal expected weight. It took a heating pad and 2.5 hours of massage with warmed towels before the fawn was finally able to hold up her head (fig. 10.1).

Figure 10.1. At nine days old, this fawn would search out a place with a barrier at her back, as she is doing here under a table, and rest after nursing (photo by Christian Mungall, courtesy of Kyle Wildlife, Texas, USA).

Work to establish a sucking response began. A bottle of commercial colostrum was offered every 2 hours although only perhaps 0.04 to 0.06 liter (1.5 to 2 oz) was actually swallowed. On day 2, total consumption for the day was up to 0.21 liter (7 oz), the fawn passed meconium, and she urinated. On day 4, commercial lamb milk replacer was substituted for the colostrum and given every 4 hours. Soon the midnight bottle was dropped. From 1 week until 4.75 months, 4 bottles were presented each day

Figure 10.2. Weight and nursing amounts by age from birth for a hand-raised eastern dama gazelle female born in a Texas pasture.

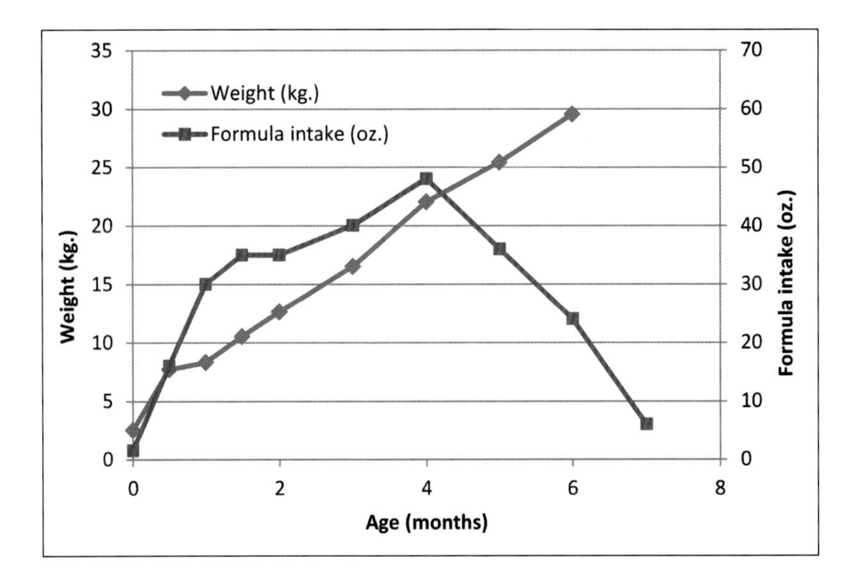

(at approximately 5:30 a.m., noon, 5:00 p.m., and 9:00 p.m.). The amount of milk offered per feeding went from 0.18 liter (6 oz.) at 1 week (the small amount was to avoid overstressing the small stomach) to 0.36 liter (12 oz) at 2.5 months. Increases were based on growth more than age (fig. 10.2).

Throughout the months of nursing, it was important to use a bottle with graduations. During the first weeks, this allowed monitoring of the amount of milk actually ingested in order to be sure that the fawn was getting nourishment. During later weeks, this allowed whoever was giving the bottle to end nursing if the fawn reached the maximum allowable for its present stage. Except for newborns, a dama gazelle mother routinely terminates nursing before her fawn loses interest. (For further discussions of developmental stages among pasture fawns, see chapters 5 and 7.)

Conspicuous changes developed as the fawn reached 1 month of age. No longer did the woman who became the fawn's surrogate mother have to go get her for nursing, open her mouth, and put in the nipple. From protruding water faucets to a dog's tail and ears, the fawn would mouth all sorts of objects. The fawn would pick up potting soil, the loose earth atop ant mounds, and pea gravel. This behavior—similar to what is seen in certain other animals such as white-tailed deer (*Odocoileus virginianus*)—may

Figure 10.3. At one month of age, the fawn was up longer after nursing and needed constant supervision. Here she removes iris flowers as both her surrogate mother and a solicitous house dog stay nearby (photo by Christian Mungall, courtesy of Kyle Wildlife, Texas, USA).

Figure 10.4. Vines (like the mustang grape here) or branches tied to the pen fence in order to give the animals a better purchase attracted instant use (photo by Christian Mungall, courtesy of Kyle Wildlife, Texas, USA).

facilitate the establishment of microorganisms in the digestive tract as needed for digestion. The dama gazelle fawn would swallow small rocks if they were not pried out of her mouth or dropped in response to a quick reprimand. The young fawn definitely needed supervision while active in artificial environments (fig. 10.3). Alfalfa in a feed tub sparked little interest, whereas the tub's rim appeared just right for chewing. The upsurge in chewing on items like sticks that the fawn was able to maneuver to the back of her mouth suggested teething. A daily ration of deer pellets was initiated, and the fawn tried to chew them but did not swallow. Thus, they were replaced with kid starter pellets, which are softer.

After seeing her companion house dog drink (a like-sized dog that always showed signs of adopting any kind of fawn or lamb brought into the house), the gazelle put her nose to water for the first time and then stepped in it. Even by 4 months—when she was still very much a nursing youngster—water use amounted only to putting her nose in and then licking the water off her muzzle.

Between 1 and 1.5 months, her companion dog, when not sent away, would still lick under the fawn's tail, eat the defecation material, and lick the ground where the fawn had just urinated. During this time, biting at plants—roses (*Rosa* sp.) were a favorite—without swallowing changed more to eating, and defecation transitioned to fecal pellets—at first light colored and then blackish like those of adults—instead of the gooey, strong-smelling meconium. After she was banished from the house and garden at 1.75 months, when pulling newspapers, iPads, and Worcestershire sauce off tables and shelves became a fascination, samples of the kinds of browse available to dama gazelles in the ranch pastures were furnished in the youngster's outdoor pen (fig. 10.4).

Particular food preferences included greenbrier (*Smilax bona-nox*, even when dry the next day), mustang grape (*Vitis candicans*, although not the immature fruits), hackberry (mainly sugar hackberry, *Celtis laevigata*), and black cherry (*Prunus serotina*). By about 2.5 months, the penchant of her species for chewing on bark and branches asserted itself. As with adult dama gazelles, this did not include chewing on bare fence posts or shed supports.

Weaning

The study female's reduced interest in nursing after 4.75 months (fig. 10.2) brought a switch to 3 bottles per day, then 2, then 1. The starter pellets were phased out in favor of regular deer pellets at 6.25 months. Milk amount went down to 0.18 liter (6 oz) by the last bottle just before 7 months. This cutoff was about 1 month later than the maximum for dama gazelles in the ranch's breeding herd. The female, now an adolescent with a goat companion, continued to be given a variety of browse samples to supplement the pan of deer pellets. At 7.5 months, these two animals were let loose during the day to forage on their own. Deer pellets like those being used to supplement the dama gazelles in the breeding herd were still provided in the pen at night.

Comparison

Reports by Douglas E. Smith of Bear Creek Exotics allow a comparison with experiences of hand raising eastern dama gazelle fawns beginning at normal weights after as much as a day of nursing naturally from their mother. Having experienced the danger to the newborns of his small species like dama gazelles when a fall "norther" blows in a quick temperature drop, Smith makes it a practice to remove for hand raising any fall or winter neonate born later than October. If first seen in the morning, the neonate gets picked up in the evening. If first seen in the evening, it gets picked up the next morning. This gives the newborn time to nurse and ingest its mother's colostrum. Smith considers fawns surviving in the pasture for this amount of time to be normal.

For the present observations, Smith started with a male fawn. Then he added a newborn female. Because the fawns had been with their mothers long enough to have the first nursing, Smith could start directly with a long-term formula mixture (Land O'Lakes commercial kid goat formula). The female fawn got her first bottles at midnight, 2:00 a.m., 7:00 a.m., 1:30 p.m., 5:00 p.m., and 1:00 a.m. By day 3, the female had learned to suck from the bottle. She took 0.09 liter (3 oz) of formula at 8:00 a.m., 0.24 liter (8 oz) at 1:00 p.m., and 0.18 liter (6 oz) at 10:30 p.m., with no midnight bottle. After 2 to 3 weeks, once she would take 0.42 liter (14 oz) dependably at each feeding, she settled into a routine of 3 feedings a day at 10:00 a.m., 3:00 p.m., and sometime between 8:00 and 10:00 p.m. This changed to 2 feedings a day at 5 weeks of age, with the formula amount increased to 0.60 liter (20 oz) per feeding. At 7 weeks, this went to 0.90 liter (30 oz) of formula twice a day. At 8 weeks, with the female nursing well, the 2 feedings finally changed to 1.05 liters (35 oz) each. By this time, the male in the same pen was also going to 2 feedings of 0.90 liter (30 oz) each and then 2 feedings of 1.05 liters (35 oz) each.

In Smith's experience, the dama gazelle fawns, like axis deer fawns (*Axis axis*) and fallow deer fawns (*Dama dama*) he has raised, did not overeat. Often, the dama gazelles seemed to want more. However, if a youngster quit nursing for as much as 30 seconds, it had had enough. It went away, and its unfinished portion was thrown out. If a youngster refused to take its bottle for 2 feedings in a row, this was a signal that it was near weaning time. At this stage, the youngster was likely to chew on its nipple at first as if unsure whether to chew or to suck. Soon, Smith would leave without offering a bottle if his charges failed to come to him when he went to feed.

After a couple of days of the male perhaps taking his morning feeding and perhaps not, the male and his female companion were reduced to just a single daily bottle of 1.05 liters (35 oz). This came at nearly 3 months for the female and about 3.5 months for the male, which was about 3 weeks older. The change might seem surprisingly abrupt except that both gazelles, now adolescents, were already regularly using their free-choice deer pellets (16 percent protein), alfalfa, and water. They had demonstrated that they could get by mainly on their pellets and alfalfa. While Smith had been away for several days, the fawns had largely refused to take a bottle from his mother. And this was in spite of their having been slower than expected to start eating pellets compared to fawns raised during colder winters that seemed stimulated to eat more. Earlier, Smith had started putting out a sweet feed mixture of grains and molasses. First, it was mixed with Calf-Manna, then not. When deer pellets were put out as well, the gazelles tended to finish their pellets but leave a little of their sweet feed. By then, it was time to discontinue the sweet feed.

Smith had been putting out alfalfa from the first week so that the fawns would become familiar with it. Eventually, they started to eat it. Then there came a day when he went out and found it gone. He also put out the small ends of mineral blocks that some of his other animals had been using.

Smith's gazelles each received their final bottle—this time with a wormer additive—at 121 days for the female (4 months of age) and 144 days for the male (close to 5 months). The male might have been weaned earlier except that it was better to wean both companions on the same day. This avoided confusion over which one was being offered a bottle and which was not. As learned at Kyle Wildlife, where the study female was being raised, 3 months is about the minimum weaning time for a dama gazelle to be able to subsist on solid food alone. As mentioned for Kyle Wildlife, 6 months is about the maximum that a mother loose on ranch pasture will let her offspring nurse. The study female had cut down substantially on her milk intake after 4.75 months of age (fig. 10.2). Thus, Smith's results are comparable to those that have been noted for dama gazelles in general. As at Kyle Wildlife, reactions of the Bear Creek youngsters rather than their age were allowed to be the major determinant of changes in their feeding regime.

Activity

The increasing activity of the fawn was also instructive to watch. At first, like a dama gazelle fawn with its real mother, the study fawn did little more than take a few steps after nursing and then collapse for her next rest period. This fawn's first steady steps came on day 4. At 1 week, the fawn could run and she would follow her human mother and especially her house dog companion. At 1 week there were little jumps, and at 2 weeks the fawn's back legs were straightening even though the pasterns, particularly in front, were still overextended. Now the fawn was staying up for about half an hour between rest periods. Nearing 1 month, the fawn had normal pasterns, she exhibited her first true stotting jump, and she would stay up for as long as 45 to 60 minutes. In another 4 days, the fawn was bouncing at her companion dog like a fawn inviting another fawn to play (without a response from the dog). Next the fawn was bouncing at "mom," with jumps back and forth as "mom" reciprocated. Soon after 1 month, another jump characteristic of dama gazelle fawns appeared. The fawn jumped up and tried to turn in the air (fig. 10.5).

This first attempt ended with the fawn coming down in a heap, but presumably she was subsequently more successful. By 2.75 months, the newly adolescent female was sometimes up for more than an hour. A mysterious coordination pattern of beating up and down with a hind leg during nursing was noticed by 3 months and dropped out at 4.5 months. (For further comments on this behavior pattern, see chapter 5.)

From the time the fawn was gaining in coordination during day 2, she demonstrated how important scent is for bonding. From then on, she would target her human mother's face, and often the mouth in particular, with her nose. When her "mom" came back after several days away, the fawn sniffed her surrogate mother's hands and legs, too, but especially her face. Twice, "mom" was able to transfer nursing duties to another person for several days in order to take a trip, but not to just anyone. The youngster

Figure 10.5. Two pasture fawns incite each other to jump in play (photo by Elizabeth Cary Mungall, courtesy of Kyle Wildlife, Texas, USA).

A

B

Figure 10.6. Fawn works to reestablish nursing: *A*, fawn blocks its mother by stepping right in front of her; *B*, fawn continues into reverse-parallel position for nursing as mother objects by wrinkling her nose and snorting (photos by Elizabeth Cary Mungall, courtesy of Kyle Wildlife, Texas, USA).

refused to cooperate with one of the ranch men, would take the bottle from another but only after going back to being caught for nursing every time, and responded well to a third man but only as long as "mom" was away. As soon as "mom" returned, that was the end of accepting a bottle from anyone else.

By week 3, as the fawn became more eager for her bottle, she started using the blocking behavior typically seen in the breeding pasture when a fawn is still focused on nursing but the mother cuts the session short by moving away (fig. 10.6).

An offspring stops its mother by cutting in front of her and sometimes rubs its head against her. In less than a week after the morning feeding was discontinued, the study female stopped showing blocking behavior or checking her caregiver's pockets for a bottle. However, she was still giving angry-sounding snorts. At 8 months of age, there was a resurgence of blocking behavior when the study fawn wanted more salted and roasted peanuts as treats. Only once, at 1 month, did the study fawn jump with her front feet on the man who had been feeding her when he cut off her nursing.

Pushing at her caregiver with her head did not start until the study female was 7.25 months old and already had horns somewhat more than 11 cm (4 in) long (fig. 10.7). Although the gesture was playful and similar to the interactions among many of the pasture youngsters, any human near her had to avoid getting butted.

Figure 10.7. By eight months of age, the surrogate mother occasionally has to fend off playful butting attempts by her growing charge (photo by Elizabeth Cary Mungall, courtesy of Kyle Wildlife, Texas, USA).

Special Behavior Needing Accommodation

During this study, several characteristics of dama gazelles became evident that anyone raising a dama gazelle would want to accommodate. Even a young fawn will run very fast. If able to get a long, straight run during play, the fawn can hit a fence and kill itself. At 8.5 months old, when let out into the pasture for the day, the growing adolescent collided with a fence that should have been familiar to her. A sudden alarm may have prompted this. One horn sheath was ripped off, leaving her with a deformed horn on that side (fig. 10.8).

Figure 10.8. Deformed right horn at 9 months after sheath was knocked off at 8.5 months (photo by Elizabeth Cary Mungall, courtesy of Kyle Wildlife, Texas, USA).

By 1 month, a fawn will no longer stay lying down when approached. This makes it hard to retrieve a fawn in order to take it back from an exercise space into a different holding area. Any consideration of where a fawn may lie should recognize that the fawn may show a great reluctance to be closed in. The growing fawn may beat its head against the door of the same dog crate that served to contain it safely at night as a newborn. After the study female was moved to the outdoor pen and given the goat companion, she was 5 months old before she would follow the example of the goat and go in out of the rain, even though their shelter was completely open on two adjacent sides.

Companionship is important for a dama gazelle. The mother is a point of security. After the trauma of being nipped on the nose when she stepped into the wrong dog's bed, the study fawn lay shivering in the lap of her human mother. The dama gazelle seeks company. This is why the goat was added after the growing dama gazelle was banished from the house to the small outdoor pen. Before the goat was added, the 1.75-month-old dama gazelle would bang on the door from the outdoor pen to the house when left alone. To do this, she had to climb the stairs from the outdoor pen to the door. Then she would be stranded and have to be carried back down the stairs. Young dama gazelles have difficulty going up or down stairs. Sometimes, the female would hazard a jump down, but a dama gazelle's legs are long and slender, and the awkwardness of her jump made it dangerous. The dama gazelle's ability to climb over a dog gate was a big surprise. The stairs had to be blocked off with a solid partition. From the goat's first appearance, the study fawn and the goat formed a close bond. They would run and jump together, eat together, and lie together at night (fig. 10.9).

A

B

Figure 10.9. Like-age three-month-old goat joins gazelle as an inseparable companion: *A*, both eat feed pellets together; *B*, both lie tightly together on the landing against the house door (photos by Kathryn Kyle, courtesy of Kyle Wildlife, Texas, USA).

Figure 10.10. Gazelle and friend staying together while let loose to forage on their own during the day (photo by Elizabeth Cary Mungall, courtesy of Kyle Wildlife, Texas, USA).

Even after 7.5 months, when allowed out of their pen on dry days, the study fawn and the goat regularly stayed within sight of each other while feeding and then drew close beside each other when resting (fig. 10.10). This mimicked the behavior of youngsters in the ranch's large breeding pasture. This close bond lasted until the study female became an adult. After that, she became indifferent to the goat or its whereabouts.

Differences from Pasture Conspecifics

As the study fawn progressed from an uncoordinated brown neonate not guaranteed to survive to a roan-and-white adolescent with horns making their first curve, she allowed her human companions a close-up look at various stages on the way to adulthood. For the most part, these observations were taken as representative for the species. Nevertheless, a few points were regarded to be more subject to differences than the rest. One obvious difference was the way in which coat color changed during the transition from fawn to adult. Hair shedding was most evident from 1.5 to especially 2 months as the fawn changed color. Interestingly, the transition stages failed to match the pattern for dama gazelles in the breeding herd from which the study fawn had come (fig. 10.11).

Presumably, the study fawn's environment of house floors and concrete as opposed to ground and grass in the pasture led to the difference in the amount and timing of hair that fell out.

A

B

Figure 10.11. Progress of change from fawn to adult coloration: *A*, change well underway for hand-raised subject at 11 weeks, but not as expected; *B*, pasture fawns showing typical intermediate stage on left and grooming on right that probably helps hasten the hair shedding that occurs at this time (photos by Elizabeth Cary Mungall, courtesy of Kyle Wildlife, Texas, USA).

In addition, there were speculations as to whether pasture youngsters of similar age were showing the same relation of body size and horn growth as the study fawn. A higher potential level of nutrition and older weaning age for the study female may have made a difference. However, the study female probably had to catch up initially as far as weight. On her birthday, she was estimated at 2 to 3 kg (5 to 6 lb). A sample of 4 newborn to 2-day-old female fawns from her pasture averaged 3.6 kg (8 lb). At 2.75 months, the study fawn weighed 15.4 kg (34 lb), and a similar female from her pasture weighed 18.6 kg (41 lb). A larger sample from the breeding pasture might have provided a more representative comparison. At 7 months, the study female looked the equal of youngsters in the breeding pasture and was possibly bigger. Table 10.1 gives measurements taken from birth through weaning during the hand-raising study for the female fawn at Kyle Wildlife.

Closing

The documentation from this Second Ark Foundation project in Ingram, Texas, should help owners in Texas and elsewhere succeed when a dama gazelle needs to be hand raised. Knowing what to expect with this species is important.

Table 10.1*A*. *Measurements in kilograms and centimeters for a hand-raised female dama gazelle, from birth through weaning and release from pen*

DATE	AGE	WEIGHT	HEIGHT (at withers)	HEART GIRTH (around chest)	NECK CIRCUMFERENCE (at patch)	HORN FRONT	HORN BACK
MM/DD/YY		kg	cm	cm	cm	cm	cm
3/2/15	Birth	2.3–2.7 estimate	—	—	—	—	—
3/16/15	2 wk	7.7	54.6	40.0	18.0	Feel bumps	Feel bumps
3/24/15	3 wk	—	—	—	—	Tiny cones	Tiny cones
3/31/15	1 mo	8.3	53.0	51.4	20.5	Tiny cones	Tiny cones
4/13/15	6 wk	10.5	61.8	49.7	20.5	0.4 (not yet visible)	0.4 (not yet visible)
4/27/15	2 mo	12.6	61.7	53.6	21.5	0.9	0.9
5/18/15	2.75 mo	15.5	64.1	59.0	20.7	0.9	0.9
5/25/15	3 mo	16.5	69.8	59.3	26.7	1.1	1.1
6/29/15	4 mo	22.0	70.6	64.3	24.5	3.3	3.3
7/19/15	4.75 mo	> 22.7 estimate	74.6	65.0	24.4	5.0 (horns show well)	5.0 (horns show well)
7/27/15	5 mo	25.4	78.6	68.5	24.9	6.1	5.1
8/24/15	6 mo	29.5 estimate	78.3	74.0	25.9	7.1 (horns curve)	9.0 (no rings yet)
9/20/15	6.75 mo	—	84.5	71.5	24.4	8.4	9.3
9/29/15	7 mo	—	86.3 (at haunches)	—	—	11.0 (from base)	—
10/8/15	7.5 mo	—	87.3	73.5	26.3	—	10.0
10/27/15	8 mo	—	87.7	73.1	25.7	9.0 right / 9.3 left (above hair)	Horns flake 1/3 up

Table 10.1*B*. *Measurements in pounds and inches for a hand-raised female dama gazelle, from birth through weaning and release from pen*

DATE	AGE	WEIGHT	HEIGHT (at withers)	HEART GIRTH (around chest)	NECK CIRCUMFERENCE (at patch)	HORN FRONT	HORN BACK
MM/DD/YY		pounds	inches	inches	inches	inches	inches
3/2/15	Birth	5–6 estimate	—	—	—	—	—
3/16/15	2 wk	16.9	21.5	15.8	7.1	Feel bumps	Feel bumps
3/24/15	3 wk	—	—	—	—	Tiny cones	Tiny cones
3/31/15	1 mo	18.3	20.9	20.2	8.1	Tiny cones	Tiny cones
4/13/15	6 wk	23.2	24.3	19.6	8.1	0.2 (not yet visible)	0.2 (not yet visible)
4/27/15	2 mo	27.8	24.3	21.1	8.5	0.3	0.3
5/18/15	2.75 mo	34.2	25.2	23.2	8.6	0.3	0.3
5/25/15	3 mo	36.4	27.5	23.4	10.5	0.4	0.4
6/29/15	4 mo	48.5	27.8	25.3	9.6	1.3	1.3
7/19/15	4.75 mo	> 50 estimate	29.4	25.6	9.6	1.9 (horns show well)	1.9 (horns show well)
7/27/15	5 mo	56.0	30.9	27.0	9.8	2.4	2.0
8/24/15	6 mo	65 esti-mate	30.8	29.2	10.2	2 13/16 (horns curve)	3 9/16 (no rings yet)
9/20/15	6.75 mo	—	33.3	28.2	9.6	3.3	3.7
9/29/15	7 mo	—	34.0 (at haunches)	—	—	4.3 (from base)	—
10/8/15	7.25 mo	—	34.4	28.9	10.4	—	3.9
10/27/15	8 mo	—	34.6	28.8	10.1	3.5 right / 3.7 left (above hair)	Horns flake 1/3 up

Section 3 History and Management at Zoos, in Parks, and on Ranches

Section 3 takes a look at dama gazelles where most of us are likely to see them—in captive facilities. This can mean zoos, wildlife parks, or ranches. It has been important for dama gazelles to be in all of these.

Zoos allow people to get close and see live examples of the animals. Walking by their exhibit, placed among many other exhibits, gives visitors a sense of the dama gazelle as a component of the diversity of life. Driving through an animal park lets people get in amid this diversity and glimpse something of the habits of the dama gazelle when out among other species. Private animal parks have staff to attend to the animals and often host researchers whose projects expand our capabilities to manage the species. Ranches are more private still. People lucky enough to visit may see dama gazelles in any setting from a small enclosure the size of a conventional zoo exhibit to a rangeland pasture of 202 to 404 ha (500 to 1,000 ac) or more. In the larger pastures, most, or even all, of the food the animals eat may be from the natural plants growing there. Similarly, a lot of their other behavior can be natural, too. Public zoos and the whole spectrum of wildlife parks create awareness of the species. This matters because all of us are more likely to care about helping the kinds of animals we know about. By contrast, the ranches offer a safe place for the animals to increase in numbers with less human intervention. Thus, public areas create people passionate in their desire to help the species escape extinction, while private areas create additional populations of the animals that constitute a safeguard against extinction.

Both the aspirations and the reality of zoos have changed substantially since the 1960s. Previously, zoos struggled to acquire examples of as many species as they could for exhibition. Since then, zoos have initiated many more collective programs that allow them to work together for conservation. Animals are exchanged for breeding, and different zoos sometimes make a commitment to work with different species in order to free a larger space for each.

Now zoos struggle with their desire to educate the public about the species on display and about work to benefit these species in their native lands. The Association of Zoos and Aquariums (AZA) recognizes zoos for their efforts to expand enclosure sizes and design enclosures to include more natural elements. Donors have responded by supporting initiatives such as enclosure expansions and chances for keepers to travel overseas to participate in conservation programs.

Conservation pays. In the United States, this realization has done much to divide the zoo community from its former partners, the wildlife ranchers like those in Texas. Zoos and wildlife ranchers used to work together to benefit from each other's resources—species and space. Then, growth of the animal rights movement encouraged people to denounce hunting of the beautiful and often vanishing species seen in zoos. However, wildlife ranches depend on hunting to give their animals economic value. The animals are bought and sold to bring in the money necessary for their upkeep. Ranches have no entrance fees and no outside donors. Also, animals loose in rangeland pastures tend to increase to numbers higher than their pastures can support. This means that herds have to be culled. Trophy hunting takes only a few. Live sale for breeding stock and culling for venison are used when more animals need to be removed.

By now, many people associated with zoos and drive-through parks have decided that it is necessary to mend relations with wildlife ranchers. In today's world of vanishing species, zoos can no longer count on ordering replacement animals from the wild. In addition, various methods for reducing reproduction in zoos in order to keep within space restrictions have been less successful than anticipated. For instance, females put on chemical birth control may not come back into breeding condition once administration stops. Ranches can supply animals and can buy surplus. This lets zoo females continue to breed as they are adapted to do. The new drive to implement a Source Population Alliance with members from wildlife ranches as well as from zoos and wildlife parks seeks to give all members a way to work together to keep their species alive for the indefinite future as well as to permit the continued lifestyle of each of their partners.

US wildlife ranches, like those so common in parts of Texas, have an additional set of issues to deal with in continuing to raise exotics—"exotics" being animals native to other parts of the world. These are usually hoofed animals like deer, antelope, wild sheep, and wild goats. With human populations rising and undeveloped land diminishing, the numbers of more and more of these sorts of animals are declining in their original homelands. Species are becoming increasingly likely to join the dama gazelle in being listed under the US Endangered Species Act. Although the act was instituted to save species from extinction—such as the wealth of native US species that are held as a public trust rather than being privately owned—government delays, permits, and red tape have done just the opposite when confronting private animal owners.

Because foreign species as well as US natives are on this list, private animal owners must comply with the regulations. The dama gazelle has a special exemption that allows, without permit, the principal kinds of management—including hunting, transporting, buying, and selling—that ranchers need in order to maintain dama gazelle herds. When the exemption was removed during a time when courts were debating challenges to its legality, the consequences were clear. Many private owners stopped working with the dama gazelle plus the other two animals being debated (addax, *Addax nasomaculatus*, and scimitar-horned oryx, *Oryx dammah*). Under the full provisions of the Endangered Species Act, numbers dropped instead of rising. After the exemption was returned, numbers started coming back up. Thus, today's situation has been a boon for dama gazelles, but it depends on people in the political and legislative spheres understanding the market forces that allow private owners to fund wildlife projects that see no public dollars.

In Europe, the major issue has been keeping the species alive. Because western dama gazelles were doing better than eastern ones in European collections, the zoo associations from Europe and North America made a pact. European zoos would concentrate on western dama gazelles, while North American zoos would concentrate on eastern dama gazelles. It is the eastern dama gazelles that have been doing so well on ranches. Western dama gazelle stock in the United States was gathered together on a Texas ranch in hopes that this subspecies would start doing as well as the eastern animals. In spite of the animals being split onto two ranches, the desired increases did not materialize. Loss of viability as a result of low variability among the western animals has been suspected. Now all are back on the first ranch with a change of breeding male, and reproduction has improved again. However, the future remains uncertain. Because the western dama gazelle has gone extinct in the wild, there are no other sources to draw on. Nevertheless, the original Spanish rescue center and European zoos to which animals had been dispersed have been able to send stock to breeding centers in and near native African habitat.

For the most part, collections on the Arabian Peninsula are poorly known outside the Middle East. Unlike the zoos on other continents, Arabian collections sometimes keep both eastern and western dama gazelles. An intriguing possibility for conservationists is that a few dama gazelles unrelated to any of the animals known elsewhere might be in private hands in Arabia.

Captive dama gazelles on four continents all contribute to conservation opportunities. Whether in a zoo, in a park, or on a ranch, dama gazelles in all types of captive facilities increase the chances for work that will keep the species alive.

Chapter 11

North American Zoos

Rise and Fall—and Rise Again—of an Iconic Desert Species

Adam Eyres

A brief history of dama gazelles in North American zoos follows—but we have to start with an understanding of the zoo world. Traditionally, zoos were menageries. Often, what was considered the best zoo was the zoo with the most species, even if there were only one or two individuals of each kind. This mentality existed for the first 70 to 80 years of zoos in the United States. Then, in the 1970s and 1980s, the push to incorporate conservation began, and it has become a major role of zoos. Small groups of scientists came together as experts on a particular species and formed Species Survival Plans (SSPs). Many SSPs of the Association of Zoos and Aquariums (AZA) began in the 1980s.

These SSPs were initially designed to maintain as much genetic diversity as possible within species, arranging for propagation with the best possible matching of animals within these small populations. Zoos understood the restriction of space within their facilities and recognized that to be successful with these limited numbers, pedigrees would have to be tracked in order to avoid as much inbreeding as possible. An SSP considers a wild-caught animal a "founder" and then makes the assumption—sometimes a very big assumption—that all wild-caught animals are unrelated to each other, provided there is no definite information to the contrary. From this point on, it is relatively easy to keep track of offspring and, through a computer model, to establish the best pairings of animals based on their relationship to one another. Aiding these efforts is a studbook established for addra (dama gazelles of the eastern subspecies *Nanger dama ruficollis*) held by the major zoos and similar institutions in North America. For some 30 years, the zoo community has worked within the parameters of the SSPs. Although the populations are still very small (only about 150 individual dama gazelles within AZA institutions in North America in 2015), the genetic diversity is as good as can be achieved with the small number of founders represented today. An analysis put out in 2012 calculated 12 known founders for the AZA dama gazelle population out of the 5 males and 15 females originally imported to the United States from Chad in 1967 (Petric and Spevak 2012).

In addition to conservation, another very important role that zoos play is education. Nearly 200 million people in the United States visit AZA-accredited zoos each year. From signage to keeper presentations to school programs to lecture series, the opportunity to educate these visitors is a great strength of the zoo community (fig. 11.1). Through education, the general public can begin to understand what is affecting animal populations worldwide and can be inspired to help in the conservation of animal species.

The beginnings of the addra holdings in North America are summarized in the addra studbook (Petric 2012). The first dama gazelle known to have entered the United States came to the Bronx Zoo on May 25, 1925. It was a single addra male. Figure 11.2 shows this male as a subadult in October 1925. He died on June 11, 1937, never having been in a breeding situation but proving that his kind could be taken care of in captivity. His age of approximately 12.5 years is old for many animals, but dama gazelles tend to be long lived (see chapter 7).

The Philadelphia Zoo also imported a dama gazelle in 1925. No record survives as to its sex. It came in on July 11, 1925, and died February 18, 1936. The first pair of dama gazelles was imported by the Chicago Zoological Society on June 4, 1935. However, they both died within six months of arrival,

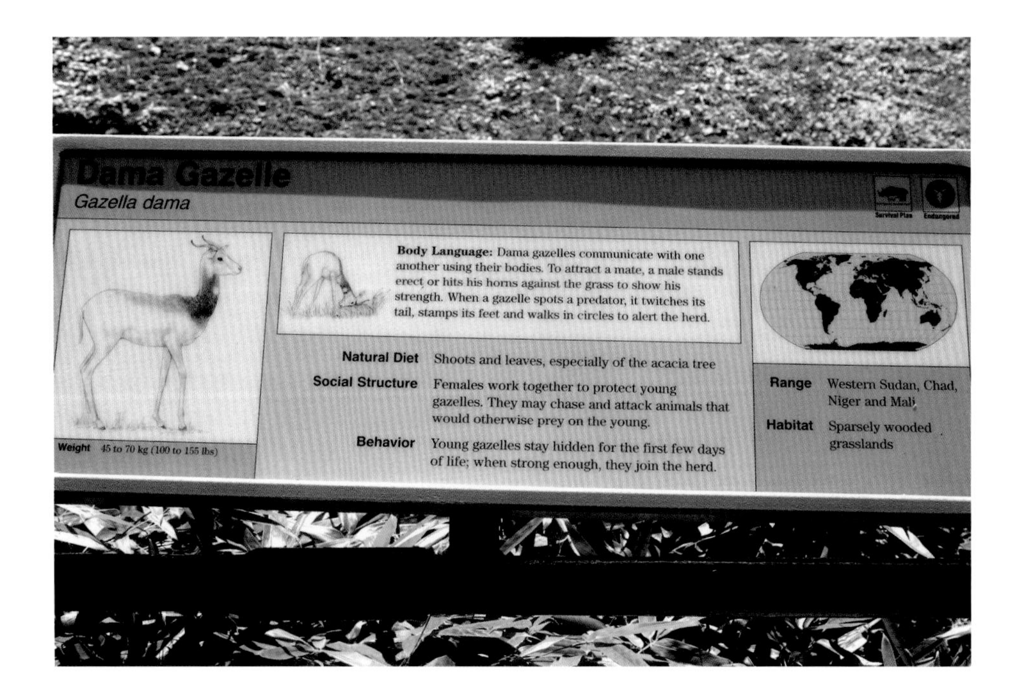

Figure 11.1. Dama gazelle sign typical of today's zoo graphics (photo by Elizabeth Cary Mungall, courtesy of the National Zoological Park, Washington, DC).

Figure 11.2. The first dama gazelle known to have reached the United States came to the Bronx Zoo as a young addra male, photographed here as a subadult in October 1925 (copyright © Wildlife Conservation Society, reproduced by permission of the WCS Archives).

Figure 11.3. Male in dama gazelle group on exhibit at the San Antonio Zoo in the 1970s courts one of the original females imported by Frans van den Brink (photo by Elizabeth Cary Mungall, courtesy of San Antonio Zoo, Texas).

the female on October 28, 1935, and the male soon after, on December 1. No offspring were produced. Finally, in 1967, 22 dama gazelles (sometimes said to have been 23) captured by Frans van den Brink from the vicinity of Ouadi Haouach in Chad, North Africa, were sent to the United States (see chapter 16). Two of these animals died in transit. The others were divided between the San Antonio Zoo in Texas and the Catskill Game Farm in upstate New York.

Most of the van den Brink dama gazelles went to San Antonio, arriving August 10, 1967, where 12 individuals (3 males and 9 females) were received (Roney 1978). Here, the 9 females were divided up into three separate breeding groups, each with its individual herd male (fig. 11.3).

The Catskill Game Farm received 2 males and 6 females. Discrepancies in the numbers remain because of a mismatch among reports about the capture and subsequent division. A letter from Frans van den Brink (published in the 2012 North American studbook, Petric 2012) says that 23 dama gazelles were captured, of which 17 went to the San Antonio Zoo and 6 went to Roland Lindeman of the Catskill Game Farm. Records for the San Antonio Zoo had to be re-created after the unexpected death of Fred Stark, who had been director when the San Antonio animals arrived (Roney 1978). Today's computer data used by the dama gazelle studbook keepers document 5 males and 15 females constituting the original "founders" of the North American captive population of dama gazelles (Petric 2012).

These two groups, San Antonio and Catskill, became the founders of all the dama gazelles subsequently kept in North American facilities, both zoos and private properties. Only years later did a very few mhorr gazelles (*Nanger dama mhorr*) appear in North America. In 1980, when Almería started to disperse some of its mhorr gazelle stock as a safeguard against heavy loss if some misfortune should strike its rescue center, the San Diego Zoo in California received 4

(Benirschke 1986). Of 12 offspring from these 4, only 7 lived, a disappointing start for the zoo's new herd (Benirschke 1986). Busch Gardens Tampa Bay in Florida exhibited a group of mhorr gazelles in the 1990s. In the mid-1990s, there was a move to gather all of the small number of mhorr gazelles in the United States to form a breeding population at Safari Enterprises in Boerne, Texas (fig. 11.4).

This was done in the hopes that, in North America, the mhorr gazelle could be positioned to flourish as the addra had on Texas ranches (for more on ranches, see chapter 15). At first, this mhorr gazelle herd increased well, but then reproduction lagged. Subsequently, the mhorr gazelle herd was split between Boerne and another Texas facility so that a single catastrophe would be unlikely to wipe out all the mhorr gazelles in North America. Now the mhorr gazelles in Texas have been brought back together at the initial ranch. Here, a change of breeding male has been followed by improved reproduction.

In the meantime, dama gazelles have continued their ups and downs as a valued component of zoo collections. Figure 11.5 shows the numbers of dama gazelles in institutions reporting to the International

Figure 11.4. Part of a breeding group after the small number of mhorr gazelles in North America was gathered together in Texas (photo by Elizabeth Cary Mungall, courtesy of Safari Enterprises, Texas).

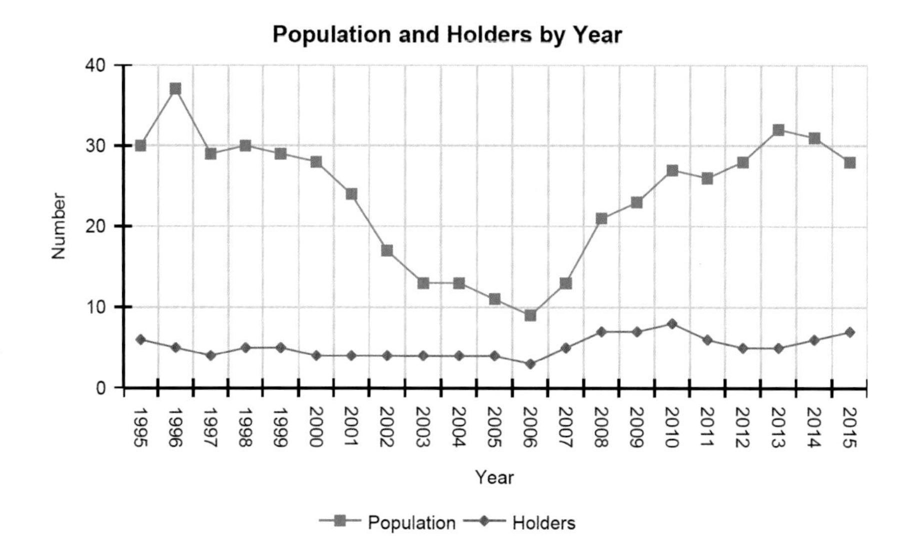

Figure 11.5. Numbers of animals and numbers of collections reporting dama gazelle holdings to the International Species Information System, an animal inventory system widely subscribed to by zoos and similar institutions, February 17, 1995, to February 17, 2015 (courtesy of Species360, www.species360.org, formerly ISIS).

Species Information System (formerly ISIS, now Species360). One factor contributing to the dip in numbers may be the AZA decision for North American facilities to focus on the eastern dama gazelle (addra) and to let European facilities focus on the western dama gazelle (mhorr gazelle).

A 2009 Regional Collection Plan clearly states this policy. Nevertheless, this fall and rise may be more reflective of the nature of working with restricted space. As the population of a species increases and available space fills, breeding is stopped. Then, after dispositions and deaths make space available again, breeding is encouraged. Thus, animal numbers can show a pronounced fluctuation while the number of holders remains relatively constant.

It is impossible to focus only on public zoos when looking at the history of dama gazelles in North America. From the very beginning, the Catskill Game Farm was a private facility, although it operated like a zoo and was open to the public. From its conception, the Catskill Game Farm was unrelated to the American Association of Zoological Parks and Aquariums (AAZPA), precursor to the AZA. Some of the biggest participants in dama gazelle conservation have been private holders. In addition to the Catskill Game Farm in New York State, there is White Oak Conservation Holdings in Florida. The successful breeding at the San Antonio Zoo and at the Catskill Game Farm created animals that became available for other owners, who then continued to grow the populations. Some of the early dama gazelle surplus went to Busch Gardens Tampa Bay in Florida and to the San Diego Zoo Safari Park in California. The San Antonio Zoo also worked with many privately owned ranches in Texas and provided not only dama gazelles, but also other exotic hoofed stock (hoofed animals of species native to foreign countries) from the zoo's highly successful ungulate breeding programs. By 1974, a wider selection of facilities was receiving addra stock, including the Oklahoma City Zoological Park, which brought in six individuals to build up a herd. Visitors particularly remarked on the grass cover in their outdoor enclosure. The National Zoological Park in Washington, DC, where a herd was started in 1976, is among the zoos that have put their gazelles among trees as part of the natural vegetation (fig. 11.6).

Figure 11.6. Addra females among trees and branches in their zoo enclosure (photo by Elizabeth Cary Mungall, courtesy of the National Zoological Park, Washington, DC).

As different facilities put together herds of the eastern dama gazelle, gene flow between the breeding lines of van den Brink's San Antonio and Catskill groups started early. As a prime example, the Catskill Game Farm's facility in Ocala, Florida, sent eight dama gazelles to White Oak between 1977 and 1981, the Catskill Game Farm in New York State sent a male, and another sire was added from the San Antonio lineage (Petric 2012). From White Oak's original group plus several future acquisitions, White Oak has produced more than 300 offspring. Fossil Rim Wildlife Center, originally a private exotics ranch, acquired AZA accreditation in 1984 and also got involved with dama gazelle conservation. The original animals came from White Oak in 1988. They have produced more than 55 dama gazelles since then. All the dama gazelles produced within both AZA facilities and the private sector combine to provide increased numbers to help with the conservation of this species.

This cooperation among private and public facilities has also had ups and downs over the last 40 years. Zoos have always known that in order to conserve a species, it is imperative that the animals breed and produce offspring. The challenge has always been space—and still is. In the 1970s and 1980s, the space problem was alleviated by the cooperative efforts of the successful breeding facilities and the larger spaces provided by private ranches, many in Texas. Zoo management and staff worked closely with private owners to ensure that zoos could continue to breed ungulates and that animals not retained for zoo programs could be sold to private ranches. There were champions of this cause, zoo directors who forged alliances with the private sector for the good of the species.

Unfortunately, there were also those who did not necessarily understand, or accept, the benefits of this system for the best interests of the species. In 1990, an episode of the television program *60 Minutes* tracked some of the animals that had left zoos and not arrived at their designated destination, or had arrived there and then been hunted. This put zoos in a negative spotlight. It implied that they were interested only in making babies so that the public would visit their zoo and did not care what happened to those animals once they grew up. Consequently, many zoos opted to stop breeding some of their animals. Many ungulate species were affected by this decision. Because zoos did not need to work with the private sector anymore, their possibilities for donors would not be hurt by increasingly negative public attitudes. Two major changes came from this decision. The first was that some zoos would now work only with AZA-accredited institutions. Thus, the space issue escalated, since most zoos did not have room for any more animals. The other issue involved contraception. This became a major study in the zoo community for many species. With contraception came fewer births and less need to go outside the AZA community for animal placement. Unfortunately, it did not take long to recognize that not producing endangered species was not going to provide animals for reintroduction as population numbers continued to decline in the wild. Also, it is natural for ungulate females to be in a state of pregnancy most of their lives, and so with contraception came pathology and an unexpected decrease in production even after the contraception was terminated. Sometimes the effects lasted much longer than anticipated after contraception was stopped. And on occasion, the previously treated animals never bred again. Another population control was to keep the sexes separate. While this addressed some of the negative side effects of certain types of contraception, it continued to create unhealthy demographics for dama gazelles because the populations were not increasing as they should, and old animals were dying off without young animals to replace them. In addition, some animals that were withheld from breeding for long periods later failed to breed at all.

Today, opinion is swinging back in favor of working together for the conservation of species. There are several AZA institutions that cooperate effectively with the private sector, and a new venture known as the "Source Population Alliance" is working to re-create and improve the original cooperative efforts of the private and public sectors. Zoos have progressed in many ways over the last 40 years, but one problem that remains, and probably always will, is the limiting factor of space. Zoos have great resources as far as money and people, but space restricts their ability to manage a large enough group of animals for meaningful conservation. In contrast, many private ranches can release exotic animals like dama gazelles into enclosures of 2 to 40 ha (5 to 100 ac) or pastures of 121 to 405 ha (300 to 1,000 ac) or more. The Source Population Alliance is working to connect private owners with not only zoos but also places like the Smithsonian Conservation Biology Institute, which have impressive assets. They have laboratories, genetics research, nutrition studies, and veterinary and keeper programs. Information exchange can benefit populations of dama gazelles living in zoos as well as those kept by the private sector. Universities and private-public relationships are important for the survival of dama gazelles in captivity. Each has resources that the other can use in its work for dama gazelle conservation. The private sector can play a crucial role in dama gazelle survival, as it has in the past, by providing space, large numbers of animals, and expertise in raising species on range land, offering animals a more natural life than is possible in the confines of small enclosures. Coupling resources like these with initiatives like the Source Population Alliance can give the private sector a greater role in conservation. Zoos can supply ranch owners with lineage records and make available important breeding stock. Cooperatives like the Source Population Alliance can continue to explore ways to benefit all types of participants and to increase opportunities for more and more partners to align their efforts for the conservation of dama gazelles.

In addition, there is another, less obvious, conservation role that several dedicated institutions have played for many years as keepers of "frozen zoos." Several zoos and universities have been banking tissue, eggs, and semen from animals that have died over the decades. With the improvement in techniques for artificial insemination, in vitro fertilization, and other assisted reproduction methods such as cloning, it is theoretically possible that no species that has been frozen will ever go extinct. However, it is all too easy for people to ignore the fact that such techniques are so expensive in time, numbers of manipulations, and money that they are not really a substitute for keeping animals alive and breeding. Another negative impact of frozen zoos is that people could potentially ignore conservation of both animals and habitats, thinking, "It doesn't matter. We still have that species." Another critical issue is that some people may think that it will not matter if a particular species continues to exist if there is no "wild" to put it back into. Do species, even those living in thriving populations that may be far from their real homes, lose value if their native habitat ceases to exist? People who think this way might do well to consider the cases in which sufficient determination allowed habitat to be re-created. For example, the Chinese did this for Père David's deer reintroductions, which started in 1985, and this was after the species had been almost extinct in the wild—it was almost entirely a park deer—since 220 AD (Mungall and Sheffield 1994). If only the species is considered, there is a tendency to ignore conservation of habitats and ecosystems. Protecting wild space is often just as important. Eliminating the perceived need to conserve a species does not correlate with saving a species with all the aspects of its life.

Zoos have done a creditable job of educating the public about species and the perils that many of them face. Zoos have also recognized over the years that what is happening in the wild can be conveyed to their visitors, and many visitors are keen to do what they can to help. It is also clear that habitats are changing and diminishing, and that acquiring animals from the wild is not an option for many species, including dama gazelles. Most zoo visitors will never visit the home countries of dama gazelles in Africa. Nevertheless, through good graphics, educational presentations, and other opportunities created by zoos, visitors can learn about the plight of dama gazelles like those they see on exhibit. Even more importantly, visitors can be encouraged to discover what role they can play in helping the species survive.

Chapter 12

European Zoos

Ups and Downs of Conserving Dama Gazelles in Europe

Tania Gilbert and Gerardo Espeso Pajares

Introduction

Chapter 11 on North American zoos has very ably provided an outline of the zoo world, and, while differences do exist on the two sides of the Atlantic, both Europe and North America share a similar zoological history and many of the same challenges. Therefore, this chapter will move straight into a discussion of the history of dama gazelles in Europe.

History of the European Population

Between 1971 and 1975, 19 western dama gazelles (5 males and 14 females), known as mhorr gazelles (*Nanger dama mhorr*), were brought to the Parque de Rescate de Fauna Sahariana, now known as "La Hoya" Experimental Field Station, Estación Experimental de Zonas Áridas (EEZA), in Almería, Spain. The goal of this initiative was to establish a European-based population of the mhorr gazelle (Cano et al. 1993, Abáigar et al. 1997, Senn et al. 2014, Espeso 2015). (For further details, see chapter 17.)

The Almería group (fig. 12.1) descended from animals captured in 1958 in the Hagounia and Tichla-Bir Ganduz regions of Western Sahara (Abáigar et al. 1997, Senn et al. 2014). All of the mhorr gazelles transported to Almería except the two from the Tichla-Bir Ganduz region were derived from the private collection of Captain, later Comandante, Julián Estalayo, an officer of the Spanish Army. Some information on his herd has been gleaned from private communications from his family, discussions with Professor J. A. Valverde (a notable biologist working in the region at that time), and a book by Professor Valverde (2004), but no detailed records on the individual gazelles were kept.

Figure 12.1. Some of the mhorr gazelles in the Almería rescue center (photo by Gerardo Espeso Pajares, courtesy of Estación Experimental de Zonas Áridas, Spain).

Despite this, enough data were available to conclude that the gazelles sent to Almería were not all unrelated founders, as originally assumed. Instead, they consisted of the three wild-caught Hagounia gazelles (one male and two females) and their offspring, as they subsequently bred, plus the two young females caught in the Tichla-Bir Ganduz region and added to the project from a holding area in Villa Cisneros (Dakhla). Only four of these five wild-caught gazelles (one male and three females) produced offspring that survived to adulthood, making these four the founders of the current mhorr gazelle population, as detailed in the international studbook (Espeso 2015). Offspring of one of the Tichla-Bir Ganduz females failed to reach maturity.

Almería has supported a captive breeding program for the mhorr gazelle since 1971 (Cano et al. 1993, Abáigar et al. 1997), but it was not until 1989 that this was formalized as the Europäisches Erhaltungszuchtprogramm, now the European Endangered Species Programme (EEP) (Espeso Pajares 2008), akin to the North American Species Survival Plans (SSPs). By this time, Almería had sent mhorr gazelles to a further six European zoos as well as across the Atlantic Ocean to the San Diego Zoo in California, from which gazelles had been distributed to other zoological institutions.

Eastern dama gazelles, known as addra (*Nanger dama ruficollis*), were first officially recorded in Europe in 1971 at the Osnabrück Zoo, Germany. Data are missing on the origin of the parents, but they were probably descended from the gazelles caught in the 1967 Chad operation by Frans van den Brink (see chapter 11 for dama gazelles in the United States and chapter 16 for stock from the van den Brink capture that went to the zoos in Munich and Leipzig without circulating through the United States). Certainly, gazelles sent to the Rotterdam Zoo in 1978 originated from the San Antonio Zoo in the United States, where the foundation animals were all from the Chad capture. Thus, the European and North American addra populations are closely related (ISIS 2015) (fig. 12.2).

Figure 12.2. Female of the eastern dama gazelle, known as addra, in a European collection (photo by John Harper, courtesy of Marwell Wildlife, Winchester, Hampshire, United Kingdom).

In 1995, the Antelope and Giraffe Taxon Advisory Group (Antelope and Giraffe TAG) of the European Association of Zoos and Aquaria (EAZA) decided to phase out the small captive population of addra in Europe and focus instead on building a viable population of mhorr gazelles. Similarly, the Association of Zoos and Aquariums (AZA) based in the United States decided to phase out mhorr gazelles in favor of addra. Consequently, mhorr gazelles are managed with a coordinated breeding program in Europe, and addra in North America (Rietkerk and Glatston 2003, AZA Antelope and Giraffe Advisory Group 2008, Senn et al. 2014).

Today, there is a very small remnant European addra population of approximately 11 animals in three European zoos. There are a further 84 addra in EAZA institutions in the Middle East, plus more in non-EAZA institutions (ISIS 2015). However, legislation by the European Union aimed at the agricultural industry makes it difficult to manage the European and Middle East gazelle herds as one population. Furthermore, a number of institutions

Figure 12.3. Pair of mhorr gazelles from a group gathered into a Texas facility (photo by Christian Mungall, courtesy of Safari Enterprises, Texas).

in the Middle East hold both subspecies (Senn et al. 2014; also see chapter 13).

As mentioned in chapter 11, the few mhorr gazelles that existed in the United States were gathered together at Safari Enterprises in Boerne, Texas, and were later split for a while between two Texas properties as a precaution against unexpected loss (fig. 12.3). The addra and US population of mhorr gazelles are not part of the European coordinated breeding program, so the rest of this chapter will focus only on mhorr gazelles in Europe.

Current Population of Mhorr Gazelles in Europe

The European population of mhorr gazelles has grown slowly but steadily since 1971. Mean annual growth rate has been about 7 percent. Today, there are approximately 225 individuals in 17 institutions across seven countries: Austria, France, Germany, Hungary, the Netherlands, Portugal, and Spain. Sixteen of these institutions are zoos or safari parks each holding small mhorr gazelle groups, and one, Almería, is a research station for arid-land fauna. This research station always has been, and remains, the largest holder of mhorr gazelles in Europe. Presently, Almería maintains just under 60 percent of the European mhorr gazelle population.

The aims of the EEPs, European counterparts of the North American SSPs, are often to maximize the retention of genetic diversity, maintain demographic stability, and provide individuals for reintroduction programs, as appropriate (Ballou et al. 2010, Montgomery et al. 2010). In addition to these general aims of EEPs, the specific aims of the mhorr gazelle program are to grow the EEP population to a sustainable level of approximately 400 individuals, and to keep inbreeding from increasing above the current level.

The European mhorr gazelle population has a reasonably stable age distribution with a sex ratio slightly biased toward females (100 males:125 females) (fig. 12.4).

However, genetic diversity has eroded since the population was founded, and inbreeding levels are high (population inbreeding coefficient $F = 0.2723$) because of the low number of founders and their unequal representation in the descendant population. Inbreeding is a concern because it depresses fitness components like reproduction and longevity in naturally outbreeding species, increasing the risk of population extinction (Frankham 1995). While Ibáñez et al. (2011, 2013) found that pedigree estimates of inbreeding did not correlate with juvenile mortality or body size for mhorr gazelles, a

Figure 12.4. Mhorr gazelles get fresh fodder (photo by Gerardo Espeso Pajares, courtesy of Estación Experimental de Zonas Áridas, Spain).

previous study by Alados and Escós (1991) showed that high inbreeding coefficients were correlated with reduced longevity in dama gazelles and had a pronounced effect on their fecundity and juvenile survival. Furthermore, Ruiz-López et al. (2012) found that molecular genetic measures of inbreeding were correlated with semen quality in mhorr gazelles, a key fitness trait. The studies on dama gazelles may be contradictory, but inbreeding remains a challenge for the European mhorr gazelle population because it is small, closed, and already inbred, and it risks becoming more so unless new, unrelated founders can be added to the population.

Concurrent loss of genetic diversity may also reduce the population's adaptive potential, limiting its ability to respond to environmental change and potentially having an impact on the success of reintroduction projects (Frankham 2003, Senn et al. 2014). Added to this is the risk of genetic adaptation to captive conditions because the environment in European institutions differs substantially from the natural habitat where the mhorr gazelle evolved. Such genetic adaptations are likely to be deleterious when populations are returned to their wild environments (Frankham 2008) and must be taken into account during the captive management and reintroduction of mhorr gazelles.

Source for Reintroductions

The European mhorr gazelle population has been both the direct and indirect source for a number of reintroductions and other conservation introductions to fenced, protected areas in North and West Africa with varying degrees of success (RZSS and IUCN Antelope Specialist Group 2014, Senn et al. 2014) (for discussions of these, see chapter 18). Between 1990 and 1994, 22 mhorr gazelles were taken from Münchener Tierpark Hellabrunn (Munich), Zoologischer Garten Frankfurt (Frankfurt), and Tierpark Berlin-Friedrichsfelde (Berlin TP) in Germany, and from La Hoya Experimental Field Station (Almería) in Spain, and released into Bou Hedma National Park in Tunisia. The population at Bou Hedma totaled 28 by 2000 (Jebali and Zahzah 2013) but dwindled after 2007, and there are now only three male mhorr gazelles in a separate fenced area called Haddej within the Bou Hedma Biosphere Reserve and National Park (RZSS and IUCN Antelope Specialist Group 2014). Indications that mhorr gazelles may have been native to Tunisia are from a time long before reliable historical records (see chapters 1 and 2), so this release may be considered a conservation introduction (Abáigar et al. 1997, IUCN 2012).

Almería was also the source for the 7 gazelles released into the Guembeul Fauna Reserve in

Senegal in 1984 (Espeso 2015). Several gazelles were reportedly transferred from Guembeul to the privately owned Bandia reserve and possibly to the Fathala reserve, both in Senegal, but no gazelles are thought to be there now (RZSS and IUCN Antelope Specialist Group 2014). In 2003, 9 mhorr gazelles were transferred from the Guembeul Fauna Reserve to the Katané enclosure in the North Ferlo Fauna Reserve, also in Senegal. Currently there are approximately 15 mhorr gazelles in the Guembeul Fauna Reserve and approximately 18 in the North Ferlo Fauna Reserve, all descended from just the 7 released from Almería in 1984 (RZSS and IUCN Antelope Specialist Group 2014). It is debatable whether it was appropriate to release mhorr gazelles into Senegal because that may have been where *N. d. dama* originally replaced *N. d. mhorr* (see chapters 1 and 2). However, the species is extinct in the wild in Senegal, so the release of mhorr gazelles, even if they constitute a different subspecies, is reasonable under the circumstances (Cano et al. 1993).

Between 1994 and 1998, Berlin TP, Frankfurt, and Munich sent 21 mhorr gazelles to the Rokkein enclosure in Souss-Massa National Park, Morocco. The gazelles have since died out (Wiesner and Müller 1998, RZSS and IUCN Antelope Specialist Group 2014). Munich and Almería were also the source of mhorr gazelles transferred in 1992 to the Domaine Royal R'Mila (R'Mila Royal Reserve), near Marrakech (Espeso 2015). Further stock was brought from German zoos between 1992 and 1996 (Wiesner and Müller, cited in RZSS and IUCN Antelope Specialist Group 2014). R'Mila Royal Reserve was established mainly as an indigenous-range breeding facility where gazelles could adapt to local climatic conditions, vegetation, forage, and soils (Wiesner and Müller 1998). Among the species at R'Mila, mhorr gazelles are actually outside their native range, but they have thrived there, and in 2014 they numbered more than 150 individuals (Wiesner and Müller 1998, RZSS and IUCN Antelope Specialist Group 2014). In 2008, 16 mhorr gazelles were transferred from R'Mila to the Safia Nature Reserve near Dakhla, where the herd grew to at least 41 individuals (RZSS and IUCN Antelope Specialist Group 2014). In May 2015, 10 of these gazelles wearing GPS collars were among 24 mhorr gazelles released in the reserve (for details, see chapter 19). This represents the first release of mhorr gazelles into an unfenced locale, albeit a protected area, within the indigenous distributional range of the species.

Thus, all these initiatives have used individuals from the European population of mhorr gazelles. This population is thought to have been founded by only four animals, and with each project, the released population is assumed to experience a further genetic bottleneck. The amount of genetic diversity represented in both the European population and at each release site is likely to be extremely limited compared to that for wild dama gazelles. This is supported by Senn et al. (2014), who found that the genetic diversity in mitochondrial DNA was much lower in samples from captive European and Middle Eastern populations of mhorr and addra gazelles compared with addra and the central *N. d. dama* still in the wild. Results for zoos, parks, and ranches in the United States are just becoming available (Senn et al. 2016; also chapter 3). This low level of genetic diversity in captive groups may affect the long-term viability of all these populations unless unrelated breeding stock can be acquired elsewhere.

Given the critical situation of dama gazelles in the wild and the challenges faced by captive dama gazelle populations, the future management of dama gazelles would be best served by following an integrated, or "One Plan" approach (Byers et al. 2013). This does not necessarily mean physically combining populations from different regions. Instead, it means that management strategies need to consider the whole global population rather than concentrate on each regional population or conservation unit separately, and that conservationists should work to develop a management plan that will give the species the best overall chance of survival. With so much Sahelo-Saharan wildlife on the brink of extinction, people now have a window of opportunity to make a real difference to the conservation of the dama gazelle, one of the world's rarest mammals, before the window closes.

Chapter 13

Arabian Collections

Lisa Banfield, Hessa Al Qahtani, and Mark Craig

Figure 13.1. Addra at Al Ain Zoo (photo by Xavier Eichaker, courtesy of the Al Ain Zoo, Al Ain, United Arab Emirates).

Figure 13.2. Mhorr gazelle females standing by an adult male, *right*, at Al Ain Zoo (photo by Xavier Eichaker, courtesy of the Al Ain Zoo, Al Ain, United Arab Emirates).

Table 13.1. Number of dama gazelles reported for collections on the Arabian Peninsula

COUNTRY	*Nanger dama mhorr*	*Nanger dama ruficollis*	TOTALS
United Arab Emirates	25.41 in 3 collections $N = 76$	27.42 in 4 collections $N = 69$	$N = 145$
Qatar	—	24.23 in 1 collection $N = 47$	$N = 47$
TOTALS	25.41 in 3 collections $N = 76$	51.65 in 5 collections $N = 116$	$N = 192$ in 5 collections

There are currently five known collections, both public and private, on the Arabian Peninsula that hold dama gazelles (in the United Arab Emirates and Qatar). All five have the eastern dama gazelle, addra ("addra" being the native name used by Arabs in parts of what became Sudan) (Sclater and Thomas 1897–1898).

In contrast to the situation in the United States and Europe, three collections have both addra (*Nanger dama ruficollis*) and the western subspecies, mhorr gazelle (*Nanger dama mhorr*) (figs. 13.1 and 13.2). The totals of all these holdings are relatively modest, with 116 addra and 76 mhorr as of November 2014 (table 13.1). It is possible that more populations exist in private collections that are not publicized.

Founders of the addra population are known to have come from several US institutions including the San Antonio Zoo, San Diego Zoo Safari Park, and White Oak Conservation Holdings, among others. Therefore, as far as can be verified, all the addra in these Arabian collections are considered to stem from the original van den Brink capture from the wild in 1967. (For further information on the derivation of addra stock in the United States, see chapters 11, 14, and 16.)

There is no information regarding additional animals coming from other sources. Nevertheless, a small possibility still exists that the gene pool in one or more private Middle East collections may harbor contributions from a few animals other than those coming from either the van den Brink capture of addra from Chad or the transfer of mhorr gazelles to Almería from Western Sahara (RZSS and IUCN Antelope Specialist Group 2014). This chance is illustrated by the report of a dama gazelle caught in the 1980s from the wild on the border between Chad and Sudan and living in 1993 in the Al Wabra Zoo, Qatar (Faris al Tamimi, pers. comm. to Tim Wacher, cited in RZSS and IUCN Antelope Specialist Group 2014).

All, or at least the majority, of the mhorr gazelle founders trace back to the rescue breeding center at Almería, Spain. This is where the only mhorr gazelles recorded to have been taken out of Africa were deposited for breeding in order to keep their kind from going extinct. Since the initial Arabian populations were established, there have been a limited number of supplements, as well as exchanges between facilities within the Arabian Peninsula, for both the addra and mhorr gazelle subspecies.

Editor's Postscript

One sizable supplement to the addra population on the Arabian Peninsula demonstrates that international exchanges can be arranged at any time and can be on a relatively large scale. As recently as January 2005, 44 addra were shipped from Texas to the Middle East (Johnson and Mungall 2016). These were gathered from ranches, rather than zoos, and air-lifted roughly 12,872 km (8,000 mi) to a quarantine center before being taken to their final destination. Animals of many ages, from fawns to adults, were included. Thanks to careful planning, all arrived in the Middle East in good condition. With so many animals leaving, this shipment caused a noticeable, although temporary, dip in the addra population on Texas ranches. Ranchers were prompted to let their dama gazelle herds go in the hope that the world's growing concern with conservation would mean that offspring from the animals being relocated would become part of future projects to reinforce failing native populations or replace lost native populations.

Chapter 14

Parks and Preserves

Adam Eyres

For captive dama gazelles, a living environment intermediate between the near-freedom of ranches and the space restrictions of zoos is found in various parks and preserves. Occasional specialized preserve facilities offer spacious paddocks run by a private estate or foundation. More conspicuous are the drive-through parks with a series of pastures inhabited by combinations of species (fig. 14.1). Some of these facilities have dama gazelles. An example of each, a private preserve and a drive-through park, will be featured here.

These types of habitat—preserves and parklands—beg the question, what is big? A drive-through park is likely to be anywhere from 162 to 810 ha (400 to 2,000 ac). Within this, 6.4 to 16 km (4 to 10 mi) of roads wind through several large pastures, which can each be 12 ha to more than 81 ha (30 ac to more than 200 ac). There may also be views of small enclosures of perhaps 0.2 to 1.2 ha (0.5 to 3 ac) on the side. Even in parks, dama gazelles sometimes have one of these smaller areas to themselves rather than being in with a mixture of other kinds of animals in the large pastures.

At White Oak Conservation Holdings, a 5,061 ha (12,500 ac) preserve in northeast Florida certified by the Association of Zoos and Aquariums (AZA), there is enough room for a dama gazelle breeding herd of a single male with typically five to eight females and their young, or for a bachelor group of males, to be given a single-species grassy enclosure of 0.8 to 6.1 ha

Figure 14.1. Drive-through parks often let visitors compare species, like these dama gazelles, smaller blackbuck antelope, and large greater kudu (photo by Jan Bussey, courtesy of Fossil Rim Wildlife Center, Texas, USA).

Figure 14.2. A new mother nuzzles her youngster in the relative peace of a single-species enclosure (photo by Karen Meeks, courtesy of White Oak Conservation Holdings, Florida, USA).

Figure 14.3. Visitors to a drive-through park where multiple males live in the same pasture have the chance to watch special displays as the males interact with each other (photo by Colin and Heather Headworth, courtesy of Natural Bridge Wildlife Ranch, Texas, USA).

(2 to 15 ac) (fig. 14.2). In a zoo, 0.4 to 2 ha (1 to 5 ac) would be considered big. Then there are the Texas ranch pastures, discussed in their own chapter (chapter 15), where big is 486 to even 8,907 ha (1,200 to even 22,000 ac). Each has its own advantages and disadvantages. Parks and preserves, like those featured in this chapter, offer a compromise between the artificial confines of a zoo where keepers manage the daily routine, and the wild-type hazards of living almost as if free, where food, shelter, and predators are largely up to the animals to deal with. So the intermediate accommodations that are the focus of this chapter allow a comfortable compromise where animals can retain much of their natural behavior while still benefiting from the security of monitored enclosures, regular feeding, and veterinary care (fig. 14.3).

White Oak Conservation Holdings is a prime example of a private preserve. Its genesis had the familiar elements of visits to wild homelands, recognition of immediate dangers, and subsequent action that have prompted many people to work to save wildlife. Like the ranchers who gave the San Antonio Zoo their famous antelope collection of the 1960s and 1970s (Roney 1978), and like professor and game commission member Dr. Frank C. Hibben, who led the State of New Mexico to experiment with exotics releases (Laycock 1966), the late Howard Gilman was spurred to action when his travels showed him firsthand the declines threatening Africa's captivating array of plains game (Mungall 2007a). Returning home, he set up a series of pastures on his estate in the Florida piney woods and populated them with a multitude of species. White Oak has had the eastern dama gazelle, also known as addra, since 1978. White Oak is now under new ownership, but its tradition of fostering research to further the welfare and conservation of wildlife both in captivity and in the wild continues.

White Oak Conservation Holdings credits the rapid population growth among its dama gazelles to the attention paid to maintenance and handling techniques, predominant aspects of which include the following methods (Speeg et al. 2014). From March to October, grass in the enclosures—mainly coastal bahia (*Paspalum notatum*)—offers natural forage. The herds are fed alfalfa hay (*Medicago sativa*) at approximately 1.0 kg (2.2 lb) per adult per day and a pelleted diet (ADF 16, HMS Zoo Diets, Inc.) at approximately 1.3 kg (2.9 lb) per adult per day. Mineral blocks and water are available free choice. Moderately sized enclosures of 1 ha (2.5 ac) have 1 herd male with 6 adult females and their young. As long as growing males (as young as 4 to 6 months of age) are removed by the time the herd male starts chasing them, as many as 10 to 20 total animals can be accommodated in these breeding groups without undue aggression. After 1 to 2 days of chasing, a herd male's aggression can increase to the point where it causes injuries. A 0.2 ha (0.5 ac) enclosure has been used to hold 3 adult females and their young. For the bachelors, there is a 6 ha (15 ac) paddock with grass and trees and sometimes 9 males. Bachelors that are surplus to needs at White Oak are matched genetically with females at other facilities and then moved out for breeding. White Oak has had up to 15 males together and up to 30 individuals in a breeding group. With any more than this, health issues such as elevated parasite loads require treatment. High animal numbers plus moist ground conditions aggravate difficulties with the parasites *Haemonchus* sp. and *Trichuris* sp. and create hoof problems. Consequently, management of the dama gazelles at White Oak concentrates on parasite control, hoof care, and herd structure.

To make possible the level of care required at White Oak, handling methods have been developed to allow capturing, doctoring, and separating of individual animals (Speeg et al. 2014). Manual restraint in dark barns with earth floors is the method of choice. Inside the barns, one small room is for holding and the next for handling. Usually, feed brings the animals into a wood-walled corral within their enclosure. From here, the animals are shut into the adjoining barn. Alternatively, the group can be herded into the corral by a staff member on foot, but this escalates risk because of the flighty nature of the species. This is particularly true for the youngsters one to six months old that have not become conditioned to this procedure. Once in the dark barn, the animals typically lie down. They can then be safely restrained by the horns. For young animals whose horns still lack the species-specific double curve, special care is needed not to twist a horn such that the seal where the base of the horn sheath meets the skin is broken. Over time, many adult dama gazelles become so used to the procedure that they remain lying down when restrained, even allowing neck venipuncture or hoof trimming while lying down. With this management strategy for feeding, holding, social grouping, and veterinary care, more than 300 fawns have been born at White Oak Conservation Holdings since 1980.

Another important aspect of the activities at White Oak is studies to promote successful husbandry for the species everywhere. An illustration of one of the recent projects at White Oak involving dama gazelles is the investigation of birth sex ratios detailed in chapter 7. A 1:1 ratio of males to females is generally considered normal for mammals like dama gazelles. However, numerous facilities that keep track of births among their hoofed stock find themselves faced with a preponderance of the same sex year after year. With small groups, this can jeopardize the continued survival of a species at a particular facility. At one point, White Oak was experiencing anxiety because 9 years of the past 11 had seen mainly male births. In spite of this, the last 30 years showed a nearly 1:1 ratio (Siegel 2013). For records work of this sort as well as for veterinary investigations like those using the White Oak laboratories, a facility like White Oak Conservation Holdings is ideal. It is dedicated to conservation and breeding of imperiled species and can give investigators the assistance they need while using individually known animals in naturalistic environments where the animals can still be readily located.

Modify a place like White Oak Conservation Holdings in order to let in the public—with the advantage of gate fees—and you have a model drive-through park like the Fossil Rim Wildlife Center (fig. 14.4). Here, the animal management staff is augmented by a conservation director, and there is an ongoing research program as well. For cooperative research with scientists from outside organizations, greater emphasis is likely to be on long-term projects. In addition, there can be space-sharing programs with other facilities in which the park temporarily holds animals belonging to zoos that may not have enough room. The additional species may or may not be available for public viewing. Some of the park's own animals are in limited enclosures, while others are loose in large fields. As constituted in 2013, the dama gazelle breeding herd of 1 adult male, 7 adult females, and 2 young had 12 ha (30 ac), while the bachelor herd of 7 adult males was loose in a 16 ha (40 ac) mixed-species enclosure (Speeg et al. 2014). The whole park encompasses 749 ha (1,850 ac).

Catching animals loose in large spaces, and often mixed in with other species, can be challenging. Until one and a half to two weeks of age, a fawn is likely to stay lying down when approached and can be picked up for examination or treatment. Other than

Figure 14.4. Regardless of fences and roads, park situations can offer dama gazelles an impressive degree of freedom (photo by Elizabeth Cary Mungall, courtesy of Fossil Rim Wildlife Center, Texas, USA).

a snort or an approach when her fawn is handled by a human, the mother is unlikely to launch a defense. Later, chemical immobilization, or occasionally net "gunning," may become the capture methods of choice. Once the animal is caught, it can be penned, moved to a darkened room, and hand restrained similarly to what is done at White Oak.

Along with the extra freedom is usually a host of animals of different kinds sharing the space. This has been a problem for dama gazelles. One drive-through park responded by keeping its dama gazelles virtually by themselves, although this is an expensive use of space in a park where the public pays to see lots of animals. Another drive-through park gave up exhibiting dama gazelles because they proved poor competitors at feeding stations when they had to jockey with other species in order to get supplemental feed. An additional consideration weighed by the administration when deciding whether to work with dama gazelles includes the effects severe winter weather can have on horns and tails. Dama gazelles are not adapted for the rigors of snow and ice, even though fawns in parks and preserves can survive birth and early growth with night temperatures of 32°F (0°C) (Speeg et al. 2014). When the tail of an older animal gets too cold, the end can freeze and fall off (Mungall, pers. observation). When sections of horn cores freeze, the horn is subject to breakage at any weakened point. In either case, exhibition value is compromised even though the individual may still have great conservation value. As at zoos, the public expects drive-through parks to present good examples of each species. When good examples are presented, the guests can be inspired to learn about the animals they are seeing, to remember them, and to care about their continued future. Thus, dama gazelles in drive-through parks can be valuable ambassadors for their species as long as living conditions suit their needs.

As at the many zoos that are open to the public, drive-through parks offer people who would never otherwise have a chance to view live examples of many foreign species the opportunity to see for themselves a diversity of animal life. Like an expansion of the zoo experience, drive-through parks offer wider vistas with animals interacting more as they would if they were in the wild. Through watching, photographing, and often feeding various animals while both animals and people exercise a degree of control over their own lives and desires, visitors can come away with a closer feeling of connection to the animals. Publications such as the *Exotic Animal Field Guide* (Mungall 2007a) can point people to parks that may be available within driving distance.

At the same time that parks are seeing to the care of their animals on exhibit, these same parks can be involved with conservation programs behind the scenes. One of the new projects being explored by the Fossil Rim Wildlife Center and other members of the recent animal-keeping consortium known as C2S2, which stands for Conservation Centers for Species Survival, is the formation of a voluntary Source Population Alliance (also see chapter 11). The intent is to ensure the future of dama gazelles and similar species. A number of facilities that have the species of concern, in this case dama gazelles, would commit to keeping at least a certain number of them. Owners would be free to continue their usual round of buying, selling, trading, and otherwise changing the composition of their herds as long as they did not go below the minimum number of animals promised. This program was prompted by the realization that the target number of 200 for animals like dama gazelles, as called for in the Regional Collection Plan for institutions in the Association of Zoos and Aquariums (AZA), is not enough for continuing survival. A core population of 400 dama gazelles has also been suggested to ensure that there will always be dama gazelles in the safety of captivity, but this number has been dismissed by statisticians. Now the target is 700 animals or even more. But AZA institutions do not have 700 dama gazelles. Even if all the similar North American institutions were to commit all of their dama gazelles to the program, that would—using the July 15, 2012, total—be only 155 animals. With more than 1,000 dama gazelles now on Texas ranches, ranch participation offers an obvious answer. If parks like Fossil Rim can bring ranches together to work with both parks and zoos, then maintaining a captive North American population of at least 700 dama gazelles at all times is a reachable goal. Ranches and AZA institutions used to work well together before differing attitudes about hunting drove them apart. Now parks and zoos have an added incentive to work with ranches again. The world is getting too small not to work together.

Chapter 15

Texas Ranches

Elizabeth Cary Mungall

Particularly favorable patterns of land use in the State of Texas in the United States have led to a well-established safe haven for dama gazelles on Texas ranch land (fig. 15.1). These are almost exclusively dama gazelles of the eastern subspecies, often called addra (*Nanger dama ruficollis*). With their unique coloration on an elegant frame, they have caught the fancy of private breeders. Dama gazelles have earned a place on more ranches than any other gazelle bred in the United States, even though dama gazelles have never been among the most numerous of the foreign wildlife species in Texas.

Exotics are species of foreign origin. Although foreign species such as European fallow deer and possibly European red deer have been kept on a few US estates since at least the 1700s, when George Washington's grounds keeper despaired at the depredations of fallow deer on Mount Vernon's ornamental

Figure 15.1. Part of a breeding herd of dama gazelles in a Texas Hill Country ranch pasture mingle with a small group of exotic axis deer (photo by Elizabeth Cary Mungall, courtesy of Kyle Wildlife, Texas, USA).

plantings (Thompson 2006), the involvement of the United States in exotics on a large scale is a relatively new phenomenon. Nowhere has it grown as it has in Texas. Here, more than 90 percent of the land is privately owned, and livestock raising is a major form of land use. With a relatively warm climate and private owners able to make their own rules, adding strange and interesting exotic beasts alongside the familiar mix of cattle, sheep, goats, and native wildlife in their pastures is easy.

Since the 1920s and 1930s, various foreign species have appeared on Texas rangeland, mainly as surplus zoo stock. At first, these were just curiosities. Then, during the "seven-year drought" of the 1950s, ranchers started scrambling for any type of income they could get. Ranchers sought to stay on their land in spite of falling livestock prices, escalating feed costs, and the ever-present taxes. If exotics were going to stay, they would have to contribute toward ranch upkeep. Soon, the hunting of exotics was being promoted. The high fees charged for off-take of a select portion of trophy males has kept this activity going and grown it to industry proportions. Not all ranches with exotics have hunting, and not all ranches with exotics allow hunting of all of the exotic species they keep, but the market for this kind of hunting gives all of these exotics monetary value for buying and selling. This means that managed herds can contribute to the costs of their fencing, ranch personnel, facilities, taxes, and feed. Feed can be a huge expense because ranches have increasingly instituted feeding of their exotics to supplement the natural rangeland forage. More recently, a meat sales industry has begun. Although on a small scale, it offers owners a way to realize funds while keeping female numbers in check.

Unlike the case with zoos or drive-through parks, there is no combination of gate fees, government subsidies, and philanthropic donations to carry the costs of the exotics. There is only the money brought in by buying and selling the animals themselves. Often this is coupled with a heavy addition of funds the owners draw from other businesses or investments. These are privately owned animals maintained by each owner's private funds. Costs are heavy, but for those people with land, the allure is great. Today, there have come to be many ranches raising solely this nontraditional livestock.

Dama gazelles came into this Texas exotics industry by way of the 1967 capture effort by Frans van den Brink (see chapter 16). As recounted, three males and nine females were received by the San Antonio Zoo on August 10 of that year. That allowed the zoo to set up off-exhibit breeding herds as well as the breeding herd on public display. Reproduction in the zoo brought surplus, which went to dealers. Many of these early transfers went to other zoos (see chapter 11). Three of the first to go to a wildlife ranch were

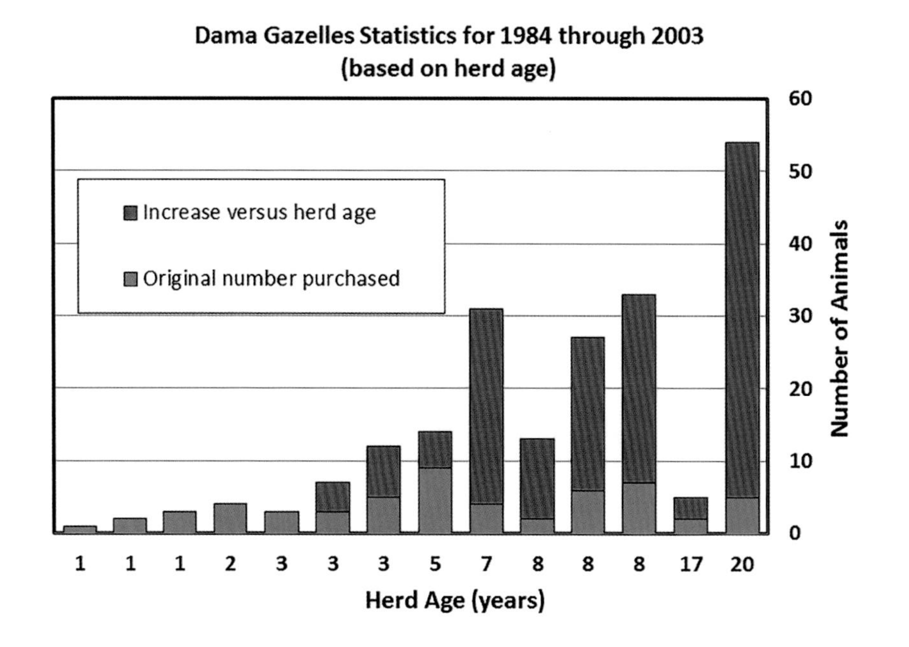

Figure 15.2. Pattern of increase in numbers of dama gazelles in Texas ranch herds from 1984 through 2003, showing a long lag time before numbers start to grow appreciably (from Mungall 2004b).

a male and two females that went to the Y.O. Ranch in the Hill Country of Central Texas on August 26, 1971 (Petric 2012). They arrived at the ranch's quarantine center as great rarities. Unfortunately, something such as a loud noise from an airplane overhead sent one of the animals crashing fatally into the fence, ending hopes of establishing a herd. It was years before the Y.O. worked with dama gazelles again.

In spite of the reputation that dama gazelles quickly earned as particularly flighty and fragile animals, their eye appeal kept ranchers trying to acquire them and trying to establish ranch populations. With their high price and the difficulty of breeding up numbers, dama gazelles—then as now—have often been a nonhunting option on exotics ranches. Statistics for 1984 through 2003 show that herds have often taken seven or eight years to become established (fig. 15.2) (Mungall 2004b). With just a pair or a trio at the start, accidents are apt to keep reproduction from taking off for years.

In the search for dama gazelle stock, some animals including in their pedigree predecessors that had gone to the Catskill Game Farm in New York State—the facility that received the other part of Frans van den Brink's dama gazelle importation—may have made their way to Texas. However, virtually all of the Texas dama gazelles—excluding a token presence of the western race, called mhorr gazelle (*Nanger dama mhorr*), which appeared very recently—probably derive from the zoo in San Antonio. The San Antonio Zoo started with more dama gazelles, had good breeding success, and was much closer to Texas properties.

The number of dama gazelles in exotics censuses by the Texas Parks and Wildlife Department, which is the state wildlife agency in Texas, and later in counts conducted for the Exotic Wildlife Association, started with a minimal representation of two in 1971 (Cook 1972) (table 15.1). Dama gazelle numbers jumped conspicuously between 1984 and 1988,

Table 15.1. Increase in census numbers of dama gazelles in Texas through 2015

YEAR	NO. OF DAMA GAZELLES	SURVEY BY (METHODS)	REFERENCE
1971	2	Texas Parks and Wildlife Dept. (Rancher and biologist reports)	Cook 1972
1979	9	Texas Parks and Wildlife Dept. (Rancher and biologist reports)	Harmel 1980
1984	14	Texas Parks and Wildlife Dept. (Rancher and biologist reports)	Traweek 1985
1988	87	Texas Parks and Wildlife Dept. (Rancher and biologist reports)	Traweek 1989
1994	149	Texas Parks and Wildlife Dept. (Rancher and biologist reports)	Traweek 1995
1996	91	Texas Agricultural Statistics Service (Mail questionnaire, phone, and visits)	Anonymous 1996
2003	369	Exotic Wildlife Association (Comprehensive phone survey)	Mungall 2004a
2010	894	Exotic Wildlife Association (Phone survey estimate*)	Anonymous 2010
2015	1,510	Exotic Wildlife Association (Comprehensive phone survey)	Meeting report 2015

Note: In 2015, owners everywhere in the United States were canvassed, but virtually all of the dama gazelles were in Texas. (The Texas Parks and Wildlife Department stopped censuses for exotics after its 1994 survey. The decrease in numbers reported for 1996 was probably due more to differences in census methods than to actual population trends.)

*Estimate based on 502 dama gazelles verified out of a potential sample of 567 ranch owners, 247 of whom could not be contacted.

jumped again between 1994 and 2003, and reached an estimated 894 by 2010 (Harmel 1980, Traweek 1985, 1989, 1995, Anonymous 1996, Mungall 2004a, 2004b, Anonymous 2010).

Interviews with state biologists for all of the state's 254 counties verified that dama gazelles are not among the few kinds of free-range exotics in Texas (Mungall 2010). Additional details taken in the 2003 survey (Mungall 2004a) reported pasture size and dama gazelle numbers for 26 of the groups. Somewhat more than half of these (54 percent) had only 1 to 10 dama gazelles, while gazelle numbers for another 31 percent of the groups were in the range of 21 to 40. The only group larger than this fell into the range of 41 to 50 animals (Mungall 2004a). By now, there are probably two ranches with somewhat more than 100 dama gazelles. The same 2003 survey found most of the Texas dama gazelles in pastures of either 2.5 to 20.2 ha (10 to 50 ac) (22 percent) or 25 to 202 ha (100 to 500 ac) (44 percent). The one really huge pasture was approximately 8,996 ha (22,220 ac). By January 31, 2015, when the Exotic Wildlife Association finished another comprehensive phone survey, 1,510 dama gazelles had been accounted for on US properties like ranches, mainly in Texas.

By way of comparison with zoo numbers, the October 2003 listing for zoos and similar major collections reporting to the International Species Information System (ISIS, now Species360) was 197 (Mungall 2004a). Considering that two Texas owners of exotic dama gazelles declined to be counted in the 2003 phone census, and that the 17 dama gazelles held by one Texas exotics owner were not tallied in the phone census because they had been reported to ISIS, there were about twice as many dama gazelles on Texas ranches in 2003 (369) as there were in zoos (197).

To be more specific about where these Texas dama gazelles live, it is mainly the Hill Country of Central Texas. This is essentially the Edwards Plateau, with its dry mosaic of oak mottes, cedar brakes, and short-grass openings (fig. 15.3). In many ways, this resembles the typical habitat for dama gazelles in their native Africa before it was taken over for domestic livestock (see chapter 20). When released in West Texas pastures like the ranch study tract of

Figure 15.3. Breeding herd of dama gazelles in a Texas Hill Country ranch pasture showing the general aspect of the vegetation (photo by Elizabeth Cary Mungall, courtesy of Kyle Wildlife, Texas, USA).

more than 8,996 ha (22,220 ac), dama gazelles have also done well. Here, the drier expanses of gravel have a sparse growth of low, spiny bushes and a few grass clumps punctuated with cactus plants and tall yuccas (fig. 15.4). In the United States, other than a few dama gazelles in the relatively mild but wet climate of Florida, dama gazelles have rarely been reported as exotics in states other than Texas.

Except where ranchers make special allowances for the requirements of dama gazelles, this species is unlikely to do well in certain regions of Texas. The Panhandle of northern Texas is exposed to winter winds that sweep south from Canada and has intensive agriculture that leaves little room for herds of exotics. As a species adapted to arid lands, dama gazelles build up dangerously high parasite loads and may develop hoof problems in the wetter climate of East Texas. In South Texas, it is predators that make it hard to establish dama gazelles. Even in the Hill Country, coyotes (fig. 15.5) can take a heavy toll on fawns and adolescents unless special measures such as electric fencing are used. Dama gazelles in West Texas also have to deal with mountain lions (fig. 15.6). However, as in the Hill Country, predator control programs that favor domestic livestock lower pressure on exotics like dama gazelles as well.

Figure 15.4. West Texas rangeland used by dama gazelles showing patches of low brush and pricklypear cactus (photo by Elizabeth Cary Mungall, courtesy of Stevens Forest Ranch, Texas, USA).

Figure 15.5. Coyote predation can take a heavy toll on young dama gazelles (photo by Susan M. Cooper of a wild coyote that strolled into a nature park in Uvalde, Texas, USA).

Figure 15.6. Mountain lion numbers are kept low in West Texas by predator control programs (photo by Susan M. Cooper, courtesy of the Arizona-Sonora Desert Museum, Tucson, Arizona, USA).

Figure 15.7. Dama gazelle pasture hit by snow and ice (photo by Scott A. Smith, courtesy of Kyle Wildlife, Texas, USA).

Figure 15.8. Shelters, heat lamps, and training help cut losses to cold weather—note feeders in front of shed (photo by Christian Mungall, courtesy of Kyle Wildlife, Texas, USA).

Figure 15.9. A crowd of dama gazelles takes shelter during cold rain, including a female nuzzling her newborn, which was born in the shelter (photo by Elizabeth Cary Mungall, courtesy of Kyle Wildlife, Texas, USA).

Next after predation as a mortality factor for dama gazelles, and more than the occasional collisions with fences, are extremes of weather. Usually, these are the storms of ice and snow that are likely to descend upon the Hill Country about every 10 years (fig. 15.7). Lacking the kind of fat deposits and thick hair that protect natives like white-tailed deer (*Odocoileus virginianus*), dama gazelles depend on their owners to bring extra feed for the added energy they need to tide them over until temperatures rise.

Ears usually come through these storms fine, but many dama gazelles lose part of their tail. As well as building shelters and providing heat lamps whenever thermometer readings start to drop into the mid-40s (°F; plus or minus 7°C) and until temperatures stay dependably above 50°F (10°C) (fig. 15.8), one ranch has worked hard to train its dama gazelles to use the shelters in cold weather (fig. 15.9).

Interestingly, the animals refused to enter when the lights were white but used the shelters well after the white heat lamps were switched to red. Attractive feed on top of extra bedding helped lure the dama gazelles into the shelters. Additional feed poured by the roadsides also helped the animals through particularly cold days (fig. 15.10).

Even after getting animals accustomed to regular ranch routines, managers have to be particularly careful with dama gazelles. Strange sights or sudden sounds can trigger a panic that may send animals running into fences (fig. 15.11).

Figure 15.10. Extra feeding to help large numbers of animals through particularly cold days: *A*, feed going out; *B*, animals coming in (photo of Scott A. Smith and of animals by Elizabeth Cary Mungall, courtesy of Kyle Wildlife, Texas, USA).

A

B

A

B

Figure 15.11. Panics can result in a broken leg or neck: *A*, a ranch herd of dama gazelles takes flight; *B*, the cause of this panic was two workers on foot near the animal's enclosure (photos by Elizabeth Cary Mungall, courtesy of Kyle Wildlife, Texas, USA).

In addition to predation, weather extremes, and fence collisions, aggression among males in smaller pastures also claims lives. Even though males kept well away from females generally live peaceably together, even 20.2 ha (50 ac) is not enough to keep males from killing each other amid the competition that develops when females—especially estrous females—are present. Hills, valleys, and stands of brush that interrupt sight lines help, but there can still be problems. This has prompted the Second Ark Foundation in Texas, in collaboration with the Exotic Wildlife Association, to initiate research on space use. Concern over this conservation problem has prompted a number of organizations to partner in supporting the study. The Hill Country Chapter and the Austin Chapter of Safari Club International have both contributed to keeping the project going, as have Bisbee's Fish and Wildlife Conservation Fund, the Fossil Rim Wildlife Center, the Dallas Safari Club, and Kevin Reid of Morani River Ranch. The ranches offering study sites have also assisted in numerous ways. In all these cases, people have made a tremendous difference.

In recent research, males in a large population in a sizable pasture where mortality from male aggression had not been found were given GPS-radio collars and tracked for a year (study by Elizabeth Cary Mungall and Susan M. Cooper) (map 15.1). Their 8,996 ha (22,220 ac) of native rangeland was assumed to be more than enough for natural behavior to express itself without causing lethal harassment. Males had room to get away from each other. Under these conditions, adult males used a home range averaging more than 1,619 ha (more than 4,000 ac) and kept "core areas"—areas where at least 50 percent of their recorded locations were concentrated—of approximately 440 ha (1,087 ac) (see also chapter 6). These core areas appeared to operate as territories (see chapter 6). Females were welcome. Subadult males sometimes came in. Other adult males stayed out. Actual contact between adult males was rare. Subadult males circulated within even more extensive home ranges averaging more than 2,834 ha (more than 7,000 ac).

The next phase of the study involved using GPS and radio-tracking collars on Hill Country subjects (fig. 15.12). Here, the main research question was how both females and males assort in a Hill Country

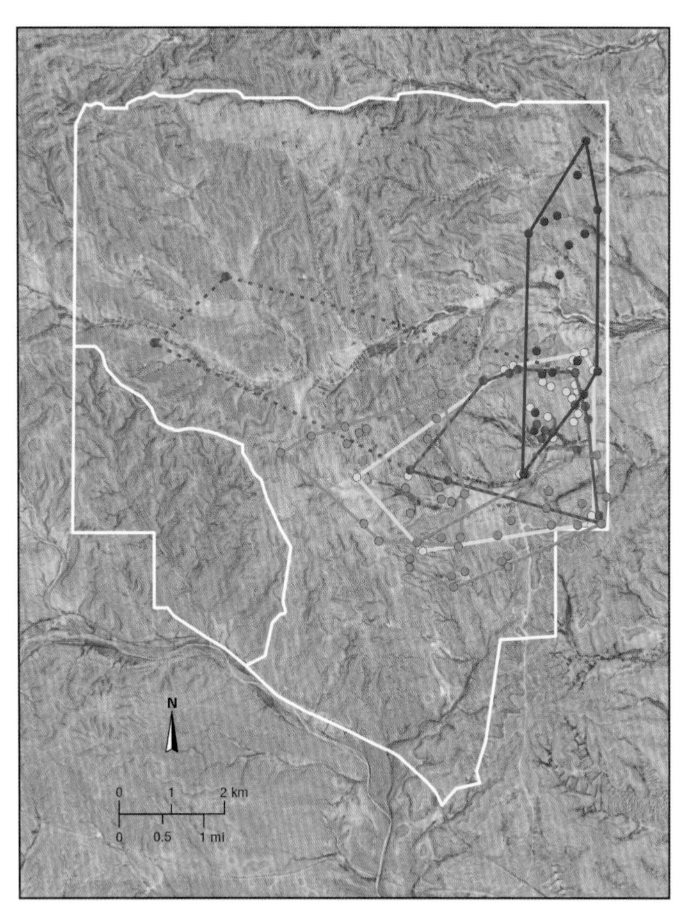

Map 15.1. Virtually nonoverlapping areas ("minimum convex polygons") used by adult males with red-striped, green, and orange collars compared with larger, mostly overlapping areas used by subadult males with purple and yellow collars during a GPS-radio collar study on a West Texas ranch (work by Mungall and Cooper, courtesy of Stevens Forest Ranch, Texas, USA).

pasture of 202 ha (500 ac), a typical size for larger Hill Country dama gazelle pastures. For part of the study these females were with male company, and for part of the time they were on their own. Then three new adult males were added to the study pasture. In summary, this continued research by Mungall and Cooper found that after the familiarization period in the pasture, the six females, one adolescent male, and young adult herd male stayed together until the death of the young adult herd male. After this, the females split into two groups, and one of the females went back and forth between groups. As soon as the new adult males were added, the young male—now a subadult—was chased away from the females. Although all the gazelles frequented all parts of the pasture enough to have the entire space considered within their home range, each animal had its core area. One new male settled into a core area of 31 ha (77 ac) where the main female group was, and another of the new males spent most of his time in the vicinity of the second female group, ending up with a three-part core area of 80 ha (197 ac). The third new male spent half his time in a core area that overlapped both of these other "properties." Like the collared males in the huge West Texas pasture, the other two new males had little contact with each other. Meanwhile, after the maturing male had been chased out of the main female group, he returned occasionally, spending part of his time alone and part of his time with one or another of the mature males.

Thus, the Hill Country dama gazelles in the smaller study pasture acted much like the West Texas dama gazelles in the huge and more open study pasture except that the Hill Country animals used brush patches, pasture shape, and steep slopes rather than expansive vistas to keep separated from each other. This shows a possibility that, at least with multiple zones out of sight of each other and with careful oversight by management—as when a maturing male may be vying for space and may need to be withdrawn for his own safety—dama gazelle groups with several males as well as numerous females can be kept together in the pasture sizes more common in the Hill Country.

Both for the sake of herds in Texas and for management or reintroduction in their African homelands, these space-use questions are important. Research in Texas is designed to assist in safeguarding the species in Texas as well as to increase the effectiveness of management and reintroductions into the wild that will keep the species from extinction.

Ranchers who own exotics are very concerned about the welfare of their animals and the prospects of the species they represent. The people managing ranches are practical stock raisers who can be counted on to be well versed in appropriate animal husbandry techniques. Owners have banded together to form associations to watch out for their interests and to go

Figure 15.12. Females and a young male, *right*, wearing GPS-radio collars in a Hill Country study (photo by Christian Mungall, courtesy of Second Ark Foundation GPS-radio collaring project at Morani River Ranch, Texas, USA).

out and find further information to fill their needs. When approached for reintroduction efforts, owners have always been willing to donate stock if the projects can go to completion. Whether it has been blackbuck antelope for Pakistan sent in the 1970s, Père David's deer to restock China offered in the 1980s (in the event, only stock from England was selected), or scimitar-horned oryx offered for Senegal when requested more recently in order to bolster genetic diversity at a reintroduction center, there have always been donation offers from Texas ranchers. The owners are sincere in their offers of stock to make the trip back to native countries in order to assist conservation projects. Dama gazelles could be next.

With its large pastures of natural vegetation and the switching of breeding stock from one ranch to another, Texas is like the "new wild" for dama gazelles and other exotics. Private owners try to assemble herds from diverse sources. Once a herd is started, the rule of thumb is to switch out the breeding male every three to four years in order to minimize inbreeding. While some herds are dispersed with the death of an owner, others go to another ranch when an owner dies, and still other herds pass from one generation to the next within the same ranch family. Thus, zoos may not be any more stable a situation for a breeding program than ranches are. Zoos can change priorities as to what species they keep, just as ranches can change the species they work with.

Zoo space is very limited compared to what can be made available by the private sector. Therefore, it can greatly benefit conservation if zoos and ranches can work together. Animals that occur in only very low numbers in the United States and need special care may do better in zoos. Animals that are increasing their populations may do well with more space and the chance to express more of their natural behavior. Ranch populations in their protected habitat can be a hedge against extinction in precarious wild situations like those facing dama gazelles. Simultaneously, ranch populations can be breeding places for animals learning life skills needed to succeed in the wild if ranches are called upon to furnish stock for reintroduction programs.

This system of safeguarded exotic populations that are also suitable and available for reintroduction shows every sign of continuing indefinitely in Texas as long as government regulations allow owners relatively unhindered use of their animals. If particular species become burdened with regulations, then owners have many other species they can switch to—and they will. This almost happened with the dama gazelle. Owners were wary when the dama gazelle was added to the species listed under the US Endangered Species Act in 2005. However, because the species is an exotic, individuals born in the United States were given a captive-bred exemption to the permit requirements typical for most listed species. Thus, buying, selling, and hunting on ranches could continue much as usual. Then court cases were brought charging that the US Fish and Wildlife Service had no right to grant such an exemption, regardless of the fact that these were not native US animals and had never been part of the wild populations dwindling in their native lands. The exemption was removed. Immediately, some owners considered dropping this species. With more than 80 kinds of exotics available in Texas, owners can disband herds of any particular species that becomes difficult. Permitting through the US Fish and Wildlife Service has been a big problem for ranches that work with listed species. There are captive-bred permits to allow management and cull-and-take permits to allow hunting. The permit process is so cumbersome that ranchers cannot necessarily count on getting the required permits in time to continue their normal ranch operations. And there is no guarantee that permits, once issued, will be renewed. In addition, the permit application is written with facilities such as zoos in mind, and so ranches can have trouble even filling it out. Fortunately for the survival of dama gazelles in Texas—and in native habitat if reintroduction is called for—legislation was finally passed that reinstated the captive-bred exemption. If it had not been, the continuation of dama gazelles as an exotic in Texas would be in jeopardy. However contrary this seems to the intent of the Endangered Species Act to keep species from going extinct, this is how the interplay of regulations and private enterprise operates in relation to exotics activity in Texas. It sounds backward, but it is the truth. Private enterprise has gone a long way toward saving the dama gazelle.

Section 4 To and from the Wild

Section 4 covers the captures that gave the world captive specimens of both the eastern and western dama gazelle. The central dama gazelle has had no such event. Consequently, it remains practically unknown. In native habitat, only four truly wild populations remain for certain, and these are precariously small. Status in one or two other areas is unsure.

Thirty-five eastern dama gazelles were captured by Dutch animal dealer Frans van den Brink in 1967 and brought out of Africa on the way to the United States and Europe. Only 2 of the 35 were lost in transit. With these 33 "founders" all, or almost all, from different herds, with the 20 US arrivals split right away between two different zoos, and with each of these males set up in a separate breeding herd, more genetic variability was maintained than for the western dama gazelles. These were rescued when extinction was imminent. Professor José Antonio Valverde, assisted by naturalist Antonio Cano, his daughter Mar Cano, and others, organized a project to transport a total of 19 of the last western dama gazelles to a specially organized breeding center at Almería in southern Spain. These were animals already in captivity in Africa. They were all descended from two small captures in 1958, a capture of one male and two females plus another capture of two females. Since one of these latter two females left no issue, all of the western dama gazelles known in the world derive from only four founders out of the five specimens originally captured.

In spite of the low starting numbers, Antonio Cano and his daughter were able to increase the holdings at Almería. To guard against any local disaster, stock was eventually dispersed to zoos across Europe. As these animals also bred, a portion joined animals from Almería to go to reintroduction centers in Africa. Animals from these centers also went to additional centers. Here they stayed until 2015, when some were finally released to see whether they could survive on their own. Although the intent had always been to return the western dama gazelles to the wild, they could not be turned loose until local conditions offered them a chance. Some were killed by wild dogs before they had dispersed from the release site. Later, evidence of poachers coincided with the disappearance of more. How the remainder have fared is impossible to know, but they definitely made a beginning.

Fortunately, there is still suitable habitat available —in terms of vegetation, topography, climate, and all the physical attributes—that could be utilized by free-ranging dama gazelles. Today, poaching and human acceptance are the main factors preventing managers from deciding to release dama gazelles into the wild. And yet, risks are also mounting for animals kept long term in the captive African breeding centers. Predators can be a problem, as can diseases and parasites, food availability, accidents, and loss of genetic diversity. No answer is easy.

In the firm belief that it is more satisfactory to help a wild remnant regain stability than to reestablish a lost population, conservationists have been trying hard to assist the last wild dama gazelles in Chad and Niger. Surveys by ground and air to monitor the situation, work by wildlife departments in range states, and information campaigns among the local people are all in progress. In addition, reintroduction of scimitar-horned oryx in Chad has increased ground support and security, which should also benefit other wildlife in the area. Bringing in eastern dama gazelles, the dama gazelle native to this region, is becoming a real possibility.

Ironically, new wells and pumping stations intended to improve the lives of the area's pastoralists have made conditions worse for the gazelles. Habitat destruction is increasing and accelerated desert encroachment threatens as the greater water availability encourages pastoralists and their livestock to settle more permanently around the water. The gazelles have been retreating to more remote and unsuitable areas where they do not always survive. Overstocking with goats, cattle, and camels in good years may bring disastrous livestock die-offs during the inevitable drought cycles. A better balance between livestock interests and wildlife would allow people and animals to forge a more sustainable relationship.

All of this has led to a perilous situation for dama gazelles. Few remain in the wild. Reintroduction is difficult. Not all reintroduction centers have had good reproduction. Despite this, however, there are still animals to work with and officials willing to see the work done.

Chapter 16

Dama Gazelles Captured in Chad

Frans M. van den Brink

Background

My 1967 capture expedition in northern Africa was searching mainly for the large Sahelo-Saharan species. One of the chief animals among these was the dama gazelle as found in the far reaches of Chad. At that time, dama gazelles were fairly common in the area where I went. The 35 dama gazelles I caught have been the seed stock for all the dama gazelles of their kind anywhere else in the world. Without the success of this capture, their race of the dama gazelle would now be almost extinct.

Preparations and Travel

I was in Chad in 1967 with my wife, Elly, my friend Gerard Schoonenberg, and his wife, Wilma. If you open a good atlas, you will see Abéché, Chad (13.49° N, 20.49° E), a small town with an airport and maybe six French people living there. Our chartered plane landed there from Holland with 11,000 kg (12 tons) of material, as in the desert there is nothing. It took me a year to prepare everything. The list of what we had with us stretched to 10 pages.

I was told by experienced animal trappers that one could not keep damas alive and acclimatize them, so I took 2,750 kg (3 tons) of alfalfa (lucerne) hay with me from Holland. Then we went by road past Biltine, Chad (14.30° N, 20.53° E), and Arada, Chad (15.00° N, 20.38° E), and reached Oum Chalouba, Chad (15.47° N, 20.40° E). I would say the distance from Abéché to Oum Chalouba was some 700 to 800 km (435 to 497 mi), and a terrible road it was. We passed holes as big as 2 by 1.5 by 2 m (6.6 by 5 by 6.6 ft.)! In Oum Chalouba was our water hole. It took about 10 hours to fill our water tank of some 2 cubic meters (2.6 cubic yards) because the water was taken up in goat skins from about 30 m (98 ft) deep.

Capture Methods

From Oum Chalouba, it was a drive in the desert. There were no signs, and no roads (fig. 16.1). We

Figure 16.1. Empty sandscape by the Wadi Wash camp (photo from Gerard Schoonenberg, courtesy of the van den Brink expedition, Chad).

Figure 16.2. Camp at Wadi Wash (photo from Gerard Schoonenberg, courtesy of the van den Brink expedition, Chad).

Figure 16.3. Chase vehicle going after a group of dama gazelles (photo from Gerard Schoonenberg, courtesy of the van den Brink expedition, Chad).

Figure 16.4. Newly caught dama gazelle on the ground between the chase vehicle and the second jeep, which has brought up a crate (photo from Gerard Schoonenberg, courtesy of the van den Brink expedition, Chad).

drove to a place called Wadi Wash, not marked on the map (fig. 16.2). All I can say is that it was in the direction of Libya. I draped the alfalfa in the acacia trees and managed to acclimatize the dama gazelles quite rapidly.

The original plan had been to capture all the animals by dart gun. A veterinarian, a human doctor, his wife, a blacksmith, and my man Friz flew out ahead of me from Holland with the 11,000 kg (12 tons) of equipment. They had with them 1,000 syringes, at least eight types of immobilizing drugs and antidotes, and a double-barreled shotgun as a capture gun. They built the camp at Wadi Wash and started work. This did not go well. When my wife and I arrived eight days later, three animals had died and the veterinarian and the doctor were not speaking to each other. I sent them home. My friends Gerard and Wilma Schoonenberg flew in instead. We took a special rope of Egyptian manufacture that I had bought in Holland, of the kind used on sailing boats even from the time of the pharaohs. We rigged it up onto the shark-fishing cane that I happened to have brought. That solved everything.

The animals were caught within a circle of some 600 to 700 km (373 to 435 mi) from Wadi Wash. We would see a group of about 20 head, cut off 1, slowly push it to a flat area, and catch it within three minutes at the most (figs. 16.3 and 16.4). My wife would knock me on the back, meaning, "Sorry, three minutes over. Leave the animal."

The three-minute rule did more than just avoid the danger of heat exhaustion. In my experience, animals chased longer than three minutes would probably be dead after 3, or 6, or even 12 months. I attribute this to thrombosis, small blood clots that would break off into the bloodstream. I once read a story about a zoo in Florida that lost 12 giraffes in one week without apparent cause. All these giraffes had been caught after chases of more than three minutes.

In the early days of the capture, we were closer to Wadi Wash. Sometimes there were only three animals or only five, and each time it was the same: cut off one from the group and try to catch it. The longer we were there, the farther away they were from our Wadi Wash camp. We could catch them only in the very early morning—say 6:30 to 10:00 a.m.—and then again in the afternoon—say 4:00 to 6:00 p.m., when it got dark. Between 10:00 a.m. and 4:00 p.m., it was unimaginably hot.

We had collapsible crates with us. When one animal was caught, we would place it in a crate under a tree and then put another collapsible crate together. When another animal was caught, we also put it under a tree—an acacia tree—and so on until the last animal.

Flying the Animals Out

For the first leg of the trip back, I arranged for a series of DC-3 charter flights from our camp to Abéché (fig. 16.5). Driving by road would have

Figure 16.5. Crated animals were loaded into a plane for the flight out of Wadi Wash (photo from Gerard Schoonenberg, courtesy of the van den Brink expedition, Chad).

been too hard on the animals. Measuring where the animals would need to go inside the plane, I discovered that our crates would be 10 cm (4 in) too high. So we cut them down. Directions for marking the landing site were to light a campfire, put up a wind flag, and make a cross with white paint over marker stones. Since we had no white paint, we took white sheets from our beds. Elly and Wilma helped the men take down our metal-pipes-and-net-wire animal enclosures at Wadi Wash for the plane trip to Abéché. Then these enclosures were put up again at an empty farm in Abéché (fig. 16.6). For the final leg home, we flew in a KLM plane (fig. 16.7).

Coat Pattern Consistency and Other Animals Caught

I think all the dama gazelles came from different groups, except perhaps a few. In my opinion, they were all of the same subspecies. Besides the 35 dama gazelles, I caught 4 addax, 43 scimitar-horned oryx, 4 cheetahs, and, in the north, a Derby eland and her male calf. All animals arrived safely at "Wormer 2," our quarantine station in Bremen (fig. 16.8).

Division and Nomenclature

The majority of the dama gazelles I captured were shipped to the United States. There, some went to

Figure 16.6. Holding pen system as used at Wadi Wash and again at Abéché (photo from Gerard Schoonenberg, courtesy of the van den Brink expedition, Chad).

Figure 16.7. KLM plane loading for flight to Europe (photo from Gerard Schoonenberg, courtesy of the van den Brink expedition, Chad).

Figure 16.8. Dama gazelles in Bremen quarantine center, West Germany (photo from Gerard Schoonenberg, courtesy of the van den Brink expedition, Chad).

Roland Lindeman at the Catskill Game Farm in upstate New York. Most went to the San Antonio Zoo in San Antonio, Texas (fig. 16.9). The 2012 addra studbook for North America (Petric 2012) accounts for 22 dama gazelles from my capture in Chad. The studbook notes that 2 of the dama gazelles died in transport, and it lists 8 (2 males and 6 females) that went to the Catskill Game Farm and 12 (3 males and 9 females) that went to the San Antonio Zoo. In addition, 2 males and 8 females went to the zoo in Munich, West Germany, and 1 male and 2 females went to the zoo in Leipzig, East Germany.

Actually, the dama gazelles that went to San Antonio were originally destined for a different US collection. However, an argument with the director over who would pay the import duties resulted in cancellation of the entire stock still on order—more than half a million dollars worth. I phoned Director Fred Stark at the San Antonio Zoo. He got 12 dama gazelles gratis in exchange for a sharing agreement. I got 12 of their fawns free of charge, and after that, as long as the originals were alive, I received title to every other of their fawns and the zoo owned the other half. Those that were not required to keep the

Figure 16.9. Female from the Chad capture standing by zoo-born dama gazelles of the next two generations at the San Antonio Zoo in the 1970s (photo by Elizabeth Cary Mungall, courtesy of the San Antonio Zoo, Texas, USA).

San Antonio herd going I was able to sell to other locations. More than 146 fawns were born during the course of this agreement.

The dama gazelles arriving at San Antonio were inventoried as the central subspecies of dama gazelle, *Gazella dama dama* (Mungall, pers. comm.). Years later, they were pronounced to have been the eastern subspecies, *Gazella dama ruficollis*, now reclassified as *Nanger dama ruficollis*. Once they were shifted to *ruficollis*, they often acquired the native name "addra," which had been used by the Arabs in Dongola and Kordofan (Sudan). That one subspecies is considered to grade into the other well west of the capture area somewhere near the Chad-Niger border.

Capture Location

The location of my Wadi Wash camp for the dama gazelle capture has been of great interest to people tracing the history of dama gazelles. As presented at an Edinburgh dama gazelle conservation workshop in 2013, the capture site was a long journey northeast of Lake Chad at Ouadi Haouach, Chad (map 16.1; 16°23'22"N, 19°37'54" E) (RZSS and IUCN Antelope Specialist Group 2014).

To find the capture vicinity on a map, start at Fort Lamy, the capital (now N'Djamena), go to Abéché, and then go north some 800 km (497 mi) to Oum Chalouba. Oum Chalouba was the nearest small city and nearest water hole. From Oum Chalouba, the expedition went some 300 km (186 mi) into the desert. Here, the only markings were numbers on a stone. Where the expedition set up camp, there was just desert and sand. In today's world of shrinking native wildlife populations, the 1967 capture of dama gazelles in Chad is becoming increasingly important for the conservation of the whole species.

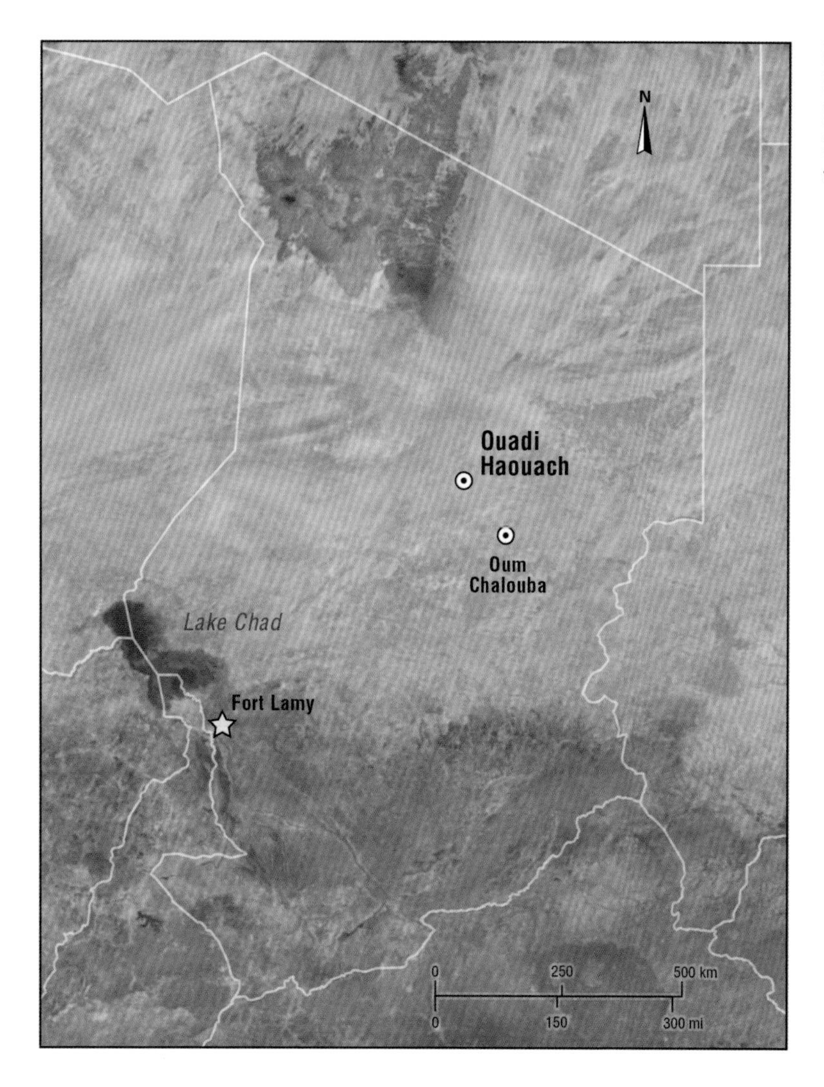

Map 16.1. Location of the van den Brink Wadi Wash capture at Ouadi Haouach, as presented at the Edinburgh 2013 dama gazelle conservation workshop.

Chapter 17

Rescuing the Mhorr Gazelle

The Path to Almería

Teresa Abáigar

The Parque de Rescate de la Fauna Sahariana (PRFS) at Almería, Spain, was created in 1971 to host a captive breeding program for the western dama gazelle (*Nanger dama mhorr*), known as the mhorr gazelle and also called mohor (fig. 17.1).

This was a risky attempt to save the last population of this dama gazelle subspecies. The plan was to gather together the animals remaining in various military barracks in the former Spanish colony of Western Sahara (map 17.1). This initiative, called

Figure 17.1. Parque de Rescate de la Fauna Sahariana (PRFS), showing the setting for mhorr gazelles at Almería, Spain (photo by T. Abáigar/CSIC, courtesy of PRFS, Almería, Spain).

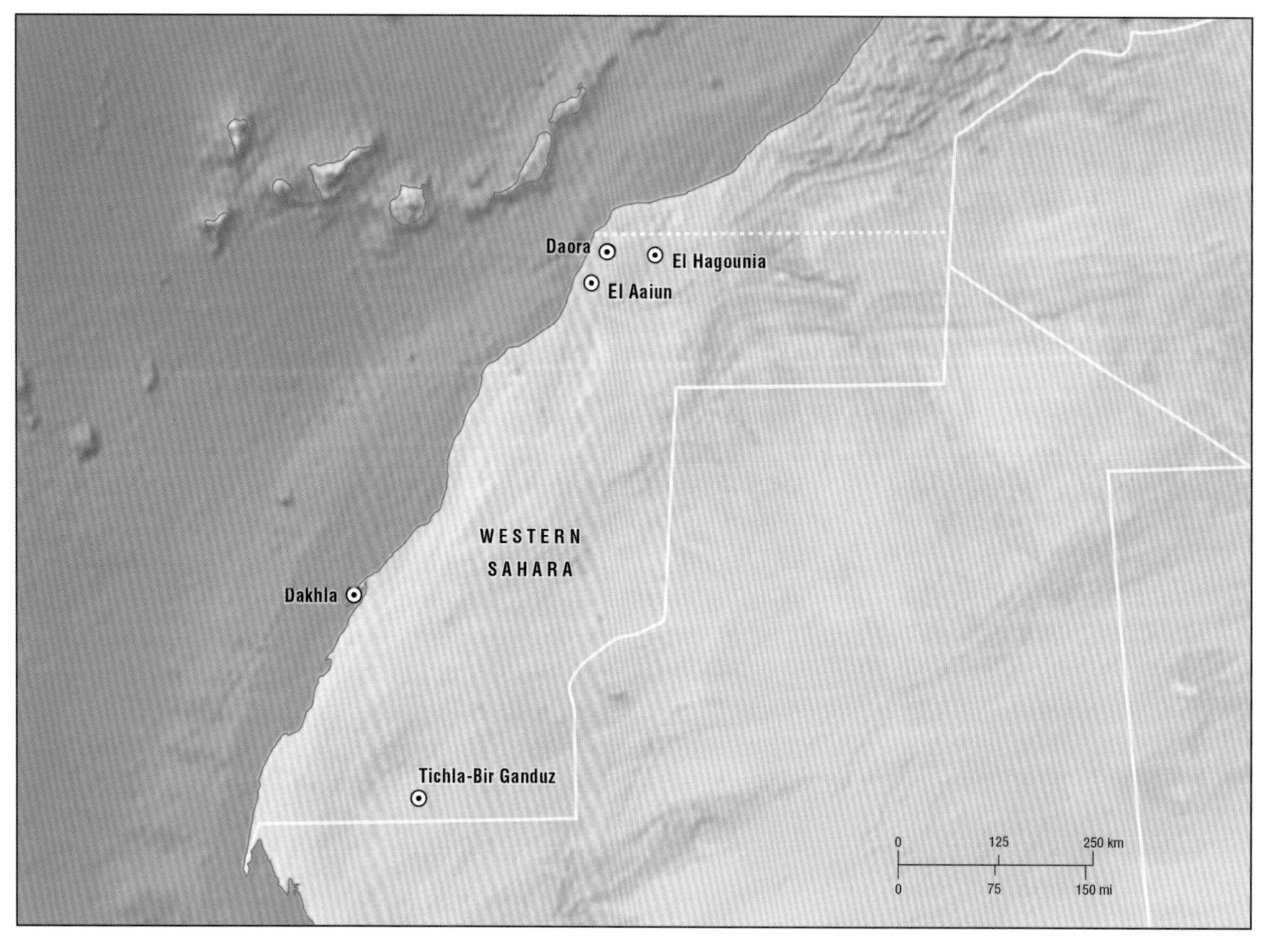

Map 17.1. Close-up of various West African locations important for the rescue of the mhorr gazelle.

"Operación Mohor," was begun by one of Spain's most distinguished ecologists, Professor José Antonio Valverde, who was responsible for the project, along with another naturalist, Antonio Cano, and his daughter Mar Cano. Ultimately, it led to the salvation of the mhorr gazelle. (For an exhaustive description of "Operación Mohor," see Valverde's 2004 memoir.)

The PRFS was established at the "La Hoya" farm, a property of the Estación Experimental de Zonas Áridas (EEZA), Instituto del Consejo Superior de Investigaciones Científicas (CSIC), located at Almería in southern Spain. In addition to the mhorr gazelle, another three species of endangered ungulates, the Cuvier's gazelle (*Gazella cuvieri*), the Saharawi dorcas gazelle (*G. dorcas neglecta*), and the Saharan aoudad, also known as the Saharan Barbary sheep (*Ammotragus lervia sahariensis*), were taken to the PRFS.

Transport of Mhorr Gazelles from Western Sahara to Almería

In 1963, in the military barracks of Daora, Comandante Julián Estalayo kept a group of 6 mhorr gazelles (1 male, 2 females, and 3 offspring). The male and 2 females for this group had been captured in 1958 from a herd located in the El Hagounia region (map 17.1). By 1970, this original group had grown to 11 individuals. By that time in 1970, Professor José Antonio Valverde, conscious that the 11 mhorr gazelles held by Comandante Estalayo were probably the last remaining population of this species, organized "Operación Mohor."

Almería was chosen as the destination for the animals obtained by the project for two main reasons. First was its climate: it was in a semidesert region considered the driest part of Europe (with less than 200 mm of precipitation per year and a mean annual temperature range of 16° to 18°C (61° to 64°F) (Alba-Sánchez et al. 2010). Second was the availability of a suitable place, "La Hoya," belonging to the EEZA/CSIC. The director of the EEZA, D. Manuel Mendizabal, accepted the creation of the PRFS at La Hoya, and Antonio Cano took on the responsibility of caring for the gazelles.

The first gazelle transport was made possible thanks to the collaboration of the Spanish Army and the Asociación para la Defensa de la Naturaleza (World Wildlife Fund–Spain). On January 13, 1971, the mhorr gazelles were anesthetized and placed in individual crates at Daora. The capture was organized by two veterinarians from the faculty of Córdoba and the veterinarian of the Córdoba zoo. The gazelles were transported by land to El Aaiun airport (60 km [37 mi] from Daora), and 7 of them (1 male and 6 females) arrived at the Almería airport on January 14, 1971, on board a DC-4 airplane belonging to the Spanish Army. The other 4 mhorr gazelles (1 male and 3 females) from the Estalayo herd were left in the care of Lieutenant Colonel Duyos in El Aaiun (Valverde 2004). Later that same year, Professor Valverde found out about the existence of 2 other mhorr gazelles, both females, in Villa Cisneros (Dakhla). These 2 had been caught, also in 1958, in the Tichla-Bir Ganduz area of Western Sahara. Again, with the help of the Spanish Army and Coca-Cola, Inc., he organized transport for these 2 females to the PRFS. They arrived in Almería on October 19, 1971.

The third and last transport of mhorr gazelles to the PRFS was in November 1975 during the last days of Spanish presence in Western Sahara. Once more, Professor Valverde, with the help of the Spanish Army, young volunteer naturalists from El Aaiun, and Mar Cano, organized transport. They picked up Duyos's group, which had grown from 4 (1 male and 3 females) in 1971 to 10 (4 males and 6 females) in 1975. This time, the gazelles were caught manually with a net before being placed in individual crates. Transport was aboard an Iberia Airlines plane that arrived at Almería on November 14, 1975.

In addition to the mhorr gazelles, they also brought other North African species. There were Cuvier's gazelles, dorcas gazelles, Saharan aoudads, fennec foxes (*Fennecus zerda*), jackals (*Canis aureus*), striped hyenas (*Hyaena hyaena*), some additional mammals, and bird species. These had originally been destined for the El Aaiun zoo, but now all stayed at the PRFS for a number of years. Eventually, some were sent to specialized collections, like birds to Granada and fennecs to the Madrid Zoo. Others, like the hyenas, lived out their lives at Almería.

Objectives of the Parque de Rescate de la Fauna Sahariana

At its beginning, the main and most urgent objective of the PRFS was to avoid complete extinction of the mhorr gazelle. Thus, all efforts were focused on ensuring the survival and reproduction in captivity of this virtually unknown species.

Eleven (2 males and 9 females) of the 19 (5 males and 14 females) mhorr gazelles originally imported to Almería have left descendants (Cano 1991, Ruiz-López et al. 2009). Nevertheless, only 4 (1 male and 3 females) of the 5 individuals originally captured from the wild (1 male and 2 females from El Hagounia plus 2 females from Tichla-Bir Ganduz) contributed successfully to the continuing population. One of the Tichla-Bir Ganduz females had 2 fawns, but the first of these 2 males died when only 2 days old, and the second died at 10 months. Then their mother died without further issue.

Thus, today's mhorr gazelle population is considered to have only 4 founders instead of 5. After more than 40 years in captivity, the survival of mhorr gazelles seems assured in captivity as long as sufficient genetic diversity can be maintained. Today, there may be more than 500 mhorr gazelles included in zoos and breeding centers scattered across Africa, Europe, the Arabian Peninsula, and North America, all of which trace back to the original group at Almería. Thanks to the vision of Professor José Antonio Valverde, Antonio Cano, and his daughter Mar Cano, and all those people who assisted with the original "Operación Mohor," the feared extinction that overtook the mhorr gazelle in the wild in 1968 (Cano 1991) has not been total extinction. Increase at Almería and in European zoos (see chapters 12 and 18) has allowed establishment of captive breeding centers in Africa, and 2015 has seen the first experimental release back into the wild (see chapter 19).

Besides the mhorr gazelle, the survival of the other three endangered North African ungulate species kept at the PRFS (Cuvier's gazelle, Saharawi dorcas gazelle, and Saharan aoudad) has been a priority within the center's objectives. All the hoofed species at the PRFS are managed according to regulations of the European Endangered Species Programme (EEP) of the European Association of Zoos and Aquaria (EAZA) and of the Association of Zoos and Aquariums (AZA) based in North America. These guidelines aim to produce demographically stable populations while retaining as much genetic variability as possible. In addition, international studbooks were established for the four species. (For additional information, see the CSIC website, http://www.eeza.csic.es/en/programa-decria.aspx.)

Reintroduction into the wild is the final objective of the PRFS. To move closer to that, the first release of mhorr gazelles back into Africa occurred in 1984 into a fenced reserve in Senegal. This was followed by other releases into fenced reserves in Morocco and Tunisia (see chapter 18 for details). Some of these releases of mhorr gazelles into reserve sites in Africa were true reintroductions, and some were outside native habitat. Reintroduction of Cuvier's gazelle occurred in Tunisia in 1999 (Abáigar et al. 2005), and the Saharawi dorcas gazelle was reintroduced in Senegal in 2007 (Abáigar et al. 2009, 2013a, 2013b).

Increasing knowledge of the species in the PRFS has been another priority because knowing details on such subjects as their biology, behavior, and ecology can promote successful survival in captivity as well as during the different phases of reintroduction. For more than 40 years, various researchers have been working on issues such as reproduction, artificial insemination, parasitology, diseases, genetics, taxonomy, and behavior.

Chapter 18

Status of Reintroductions

Teresa Abáigar

Mhorr Gazelles Start Back to Africa

Since the 1970s, when the numbers of wild dama gazelles went into precipitous decline, researchers, managers, and concerned observers have worked and dreamed of reestablishing this striking member of North Africa's fauna. Starting in 1984, dama gazelles were sent to six different African locations in three countries: Morocco, Senegal, and Tunisia (map 18.1; table 18.1). All of these dama gazelles have been of the western subspecies known as mhorr gazelle (*Nanger dama mhorr*) (fig. 18.1). No individuals of the central African dama gazelle (*N. d. dama*) are known outside Africa, so none of these can be sent back.

Figure 18.1. Prime example of a mhorr gazelle group typical of the Safia enclosure at Safia Nature Reserve (photo from CBD-Habitat Foundation, courtesy of Safia Nature Reserve, Morocco).

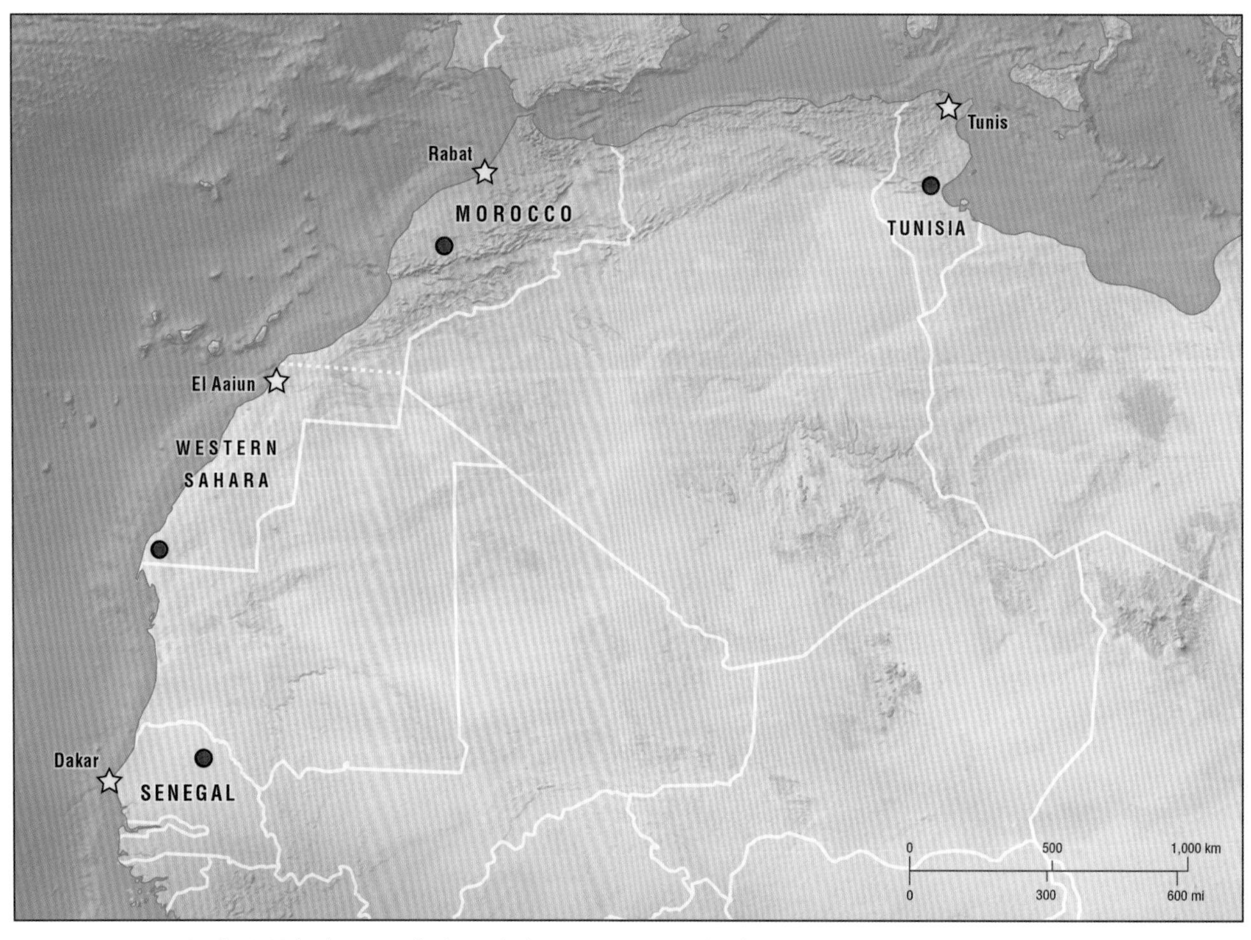

Map 18.1. Countries for initial mhorr gazelle fenced release zones in North Africa.

A few individuals have been privately held in Chad and Niger—or originate from there—mainly as pets or potential gifts, but no sources have been identified that could be drawn on for reintroduction (RZSS and IUCN Antelope Specialist Group 2014). Similarly, there could be rare cases of holdings in the Middle East or in Sudan for specimens coming from other than the well-known captures—the captures resulting in mhorr gazelles at Almería and the capture of the eastern subspecies addra (*N. d. ruficollis*) in 1967 by Frans van den Brink (RZSS and IUCN Antelope Specialist Group 2014). As an example, a wild-caught addra from the Chad-Sudan border in the 1980s was in the Al Wabra Zoo, Qatar, in 1993 (Faris al Tamimi, pers. comm. to Tim Wacher, cited in RZSS and IUCN Antelope Specialist Group 2014). However, for addra, it was the serendipitous exportation by Frans van den Brink in 1967 (see chapter 16) that gave the world numerous prospects for replacing lost populations and for bolstering the small, fragmented remaining populations of the eastern subspecies. While possible addra reintroductions are under discussion, this chapter will focus on what has happened with repatriation projects to date.

There are several reasons why only mhorr gazelles have been sent back to Africa (table 18.1). The primary reason is that the mhorr gazelle is the subspecies of dama gazelle that was taken into captivity in Europe in order to breed against the time when reintroduction would become possible. Since 1971, mhorr gazelles have reproduced well at the Parque de Rescate de la Fauna Sahariana (also called "La Hoya" Experimental Field Station, with the Spanish acronym FEH). This is a breeding center belonging to the Estación Experimental de Zonas Áridas (Consejo Superior de Investigaciones Científicas [EEZA/ CSIC]), located at Almería in southern Spain. Here, mhorr gazelles have been bred and studied. Eventu-

Table 18.1. Summary of dama gazelle reintroductions and similar releases in North and West Africa

COUNTRY Type of release	PLACE	AREA	YEAR	ORIGIN	STARTING NUMBER (males, females)	POPULATION REPORTED IN 2013
SENEGAL Outside native range	Guembeul Fauna Reserve	720 ha (1,778 ac)	1984	Almería	7 (2 M, 5 F)	13–15
Reintroduction	Katané enclosure (in North Ferlo Fauna Reserve)	440 to 640 to 1,240 ha (1,087 to 1,581 to 3063 ac) as of 2012 (of 487,000 ha [1,202,890 ac] in reserve)	2003	Guembeul	5 (2 M, 3 F)	16–18
MOROCCO Outside native range	R'Mila Royal Reserve	330 ha (815 ac) (of 465 ha [1,148 ac] in reserve)	1992 1992–1996	Almería and Munich zoo Berlin, Frankfurt zoos	6 (3 M, 3 F) < 20 added	158+
Outside native range	Souss-Massa National Park	2,000 ha (4,940 ac) Rokkein enclosure (in 33,800 ha [83,486 ac] park)	1994–1998	Berlin, Frankfurt, Munich zoos R'Mila	21 (13 M, 8 F)	0
Reintroduction	Safia enclosure (in Safia Nature Reserve)	600+ ha (1,482+ ac)	2008	R'Mila	16 (7 M, 9 F) 3 others died during transport	41
TUNISIA May constitute reintroduction but status uncertain	Bou Hedma National Park	2,000 ha (4,940 ac)	1990–1992 1994	Berlin, Frankfurt, Munich zoos Almería	8 (2 M, 6 F) of which 4 (1 M, 3 F) soon died 14 (4 M, 10 F)	3 fully adult males
TOTAL ESTIMATE						231–235+

Note: All involve mhorr gazelles after extinction in the wild.

ally, seed stock was released to several zoos in Europe (France, Germany, Italy, the Netherlands, Hungary, Spain, and United Kingdom). The relative proximity of Europe to the African homelands of this western subspecies of dama gazelle made mhorr gazelles not only available to send, but also less expensive to ship than specimens from elsewhere would have been. In addition, mhorr gazelles had completely disappeared from the wild by 1968 (Cano 1991), whereas at least a few animals of the central and eastern subspecies remain. As detailed in chapter 17 on rescuing the mhorr gazelle, it is believed that all of the currently existing mhorr gazelles in the world today are descendants from animals captured in only two areas of Western Sahara in 1958. These animals were gathered at Almería, Spain, in 1971 and 1975 (Cano 1991; Valverde 2004).

Of the six African releases of mhorr gazelles, the North Ferlo project at Katané in Senegal and the Safia project in Morocco were reintroductions, while the release in the R'Mila Royal Reserve as well as in Souss-Massa National Park (both in Morocco) were outside the known native range. Controversy about the historical presence of dama gazelles in Tunisia remains. The release at Bou Hedma in Tunisia could be considered a reintroduction (Scholte 2013). However, as pointed out in chapters 1 and 2, the original distribution of the mhorr gazelle may well have extended into Tunisia, but any native mhorr gazelles there would have died out so long ago that statements are based on conjecture.

Guembeul Fauna Reserve

First to make reintroduction hopes a reality were seven mhorr gazelles from the breeding group in Almería sent to Senegal's Guembeul Fauna Reserve (GFR) in 1984, near what was then the capital of Saint-Louis (map 18.2) (Cano et al. 1993). This

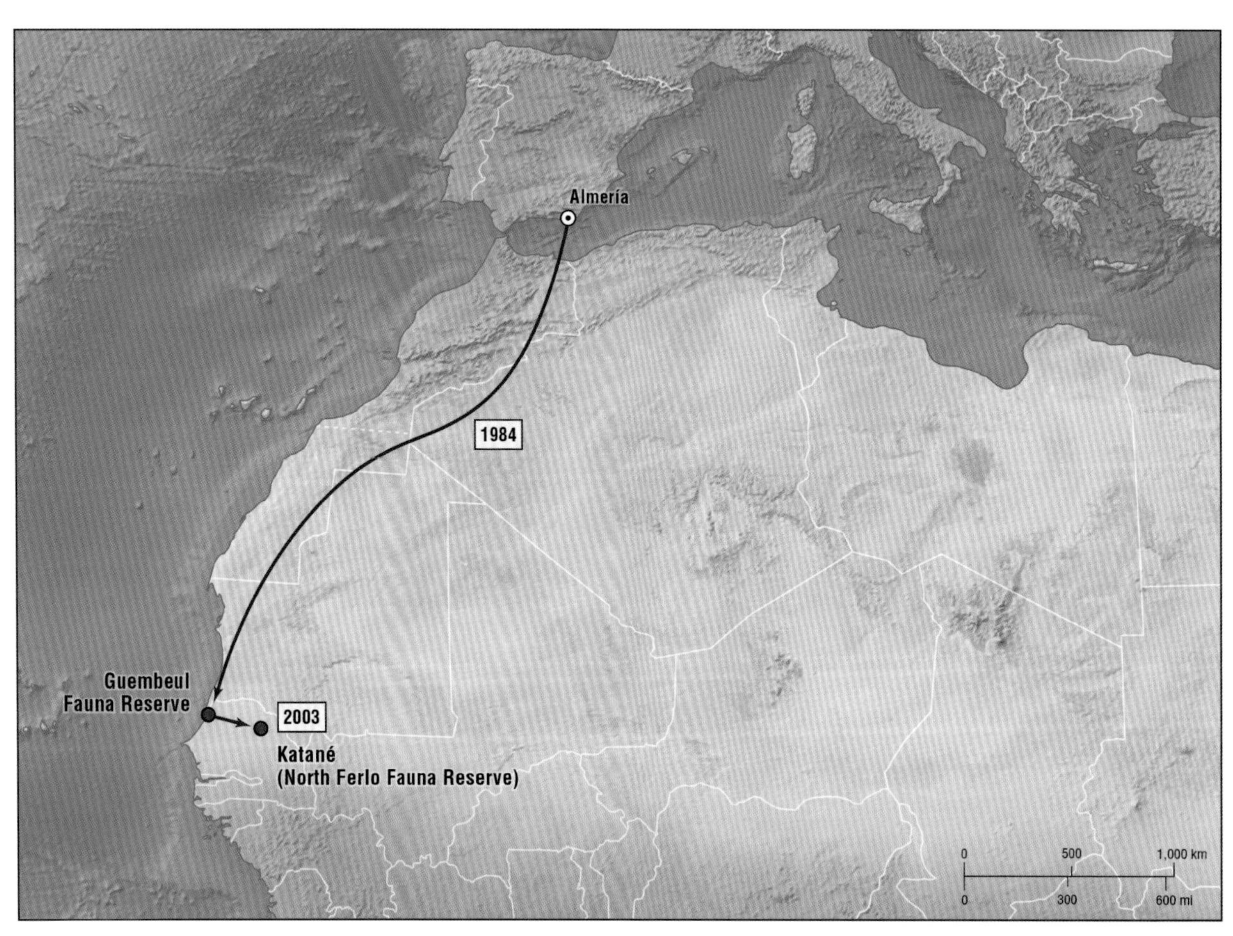

Map 18.2. Mhorr gazelle releases in Senegal, starting with Guembeul Fauna Reserve in 1984 and from there into the Katané enclosure at North Ferlo Fauna Reserve in 2003.

Figure 18.2. Vegetative cover at Guembeul Fauna Reserve, viewed toward the lake (photo by T. Abáigar/CSIC, courtesy of Guembeul Fauna Reserve, Senegal).

fenced, protected area of 720 ha (1,778 ac) near the Atlantic coast was created in 1983 with the specific purpose of protecting Sahelian ecosystem habitat and with a clear mandate to serve as a place for animal acclimatization during the first phases of reintroduction projects for Sahelo-Saharan antelope species that have disappeared from Senegal. In addition to the mhorr gazelle, these are the dorcas gazelle (*Gazella dorcas*) and the scimitar-horned oryx (*Oryx dammah*) (Abáigar et al. 2009).

Tree and shrub vegetation in the GFR is dominated by acacias (*Acacia albida*, *A. raddiana*, and *A. senegal*) and *Adansonia digitata*, *Balanites aegyptiaca*, *Capparis decidua*, *Euphorbia balsamifera*, *Prosopis* spp., *Salvadora persica*, and *Tamarindus indica* (fig. 18.2).

Cactus (*Opuntia tuna*) is progressively invading the western part of the reserve. This reduces grass cover and food availability and impedes easy access by gazelles and oryx to the acacia leaves and fruits. A third of the reserve is covered by shallow, brackish water (fig. 18.3; map 18.3).

This provides water to the animals. As the number of animals (gazelles and oryx) has increased in the reserve, supplementary food (peanut straw) and water have been provided during the last months of the dry season (March–July) before rainfall resumes. However, access to this food by mhorr gazelles is irregular because of the number of oryx and their behavior in the presence of gazelles.

The fortunes of mhorr gazelles at the GFR have gone up and down. After the initial importation of the 7 gazelles (2 males and 5 females) in 1984, the population increased to 13 (6 males and 7 females) at the end of 1992 (Cano et al. 1993). Then there is a gap of almost 10 years with no information about numbers until 2001 and 2002, when Clark (2001,

Figure 18.3. Mhorr gazelles at water's edge in Guembeul (photo by Mar Cano/CSIC, courtesy of Guembeul Fauna Reserve, Senegal).

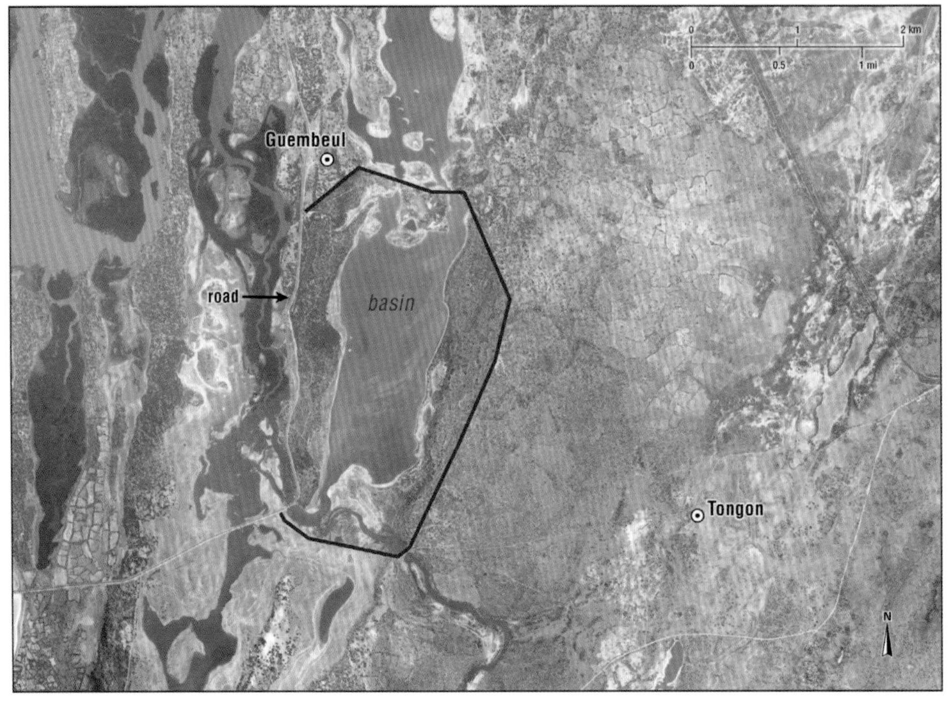

Map 18.3. Aerial view of Guembeul Fauna Reserve, Senegal, showing the lake and shoreline habitat.

2002) refers to 44 and 49 mhorr gazelles, respectively. However, these numbers seem to overestimate the population by 2002. Other observers provide a total of 32 mhorr gazelles (14 males and 18 females) in the GFR before the transport of 9 animals (2 males and 7 females) to the Katané enclosure in January 2003. In 2005, Jebali (2005) gives 20 as the total number of gazelles in the reserve. The decrease reflects the removal of the 9 individuals sent to Katané as well as the demographic disequilibrium following the transport of mhorr gazelles to Katané. If the population at the end of 2002 was 32 (14 males and 18 females), then the remaining population in the GFR after the transport of the group to Katané was 23 (12 males and 11 females). In 2007, the population increased to 28 gazelles with two well-established groups, one in the eastern part of the reserve and the other in the western part.

Then the total went down to 10 individuals (Moreno et al. 2012) because of dog attacks causing 5 deaths in 2008 (B. Youm, pers. comm.) and because of mortality within the group in the west during the capture of 2 mhorr gazelles to transport to Mauritania in 2009. As of 2013, the GFR still had 13 to 15 mhorr gazelles. The main conclusions about the development of the reintroduced mhorr gazelle population in the GFR are that (1) although numbers have risen, population trends could eventually reflect a loss of genetic diversity because of the low number of founders when the population was established and no additional gazelles being subsequently added; (2) there have been management problems for this population—mainly the mortality, sex bias, and social upheaval following captures; and (3) accidents have resulted from fence deterioration that allowed entry of dogs. These are probably the main explanations for the evolution of mhorr gazelle numbers in the Guembeul Fauna Reserve, rather than the habitat structure in the reserve, as has been claimed (Moreno et al. 2012).

Katané Enclosure in the North Ferlo Fauna Reserve

The mhorr nucleus in the GFR allowed stock to be transferred to the Katané enclosure in 2003 (map 18.4). The Katané enclosure is located in the North Ferlo Fauna Reserve (NFFR), an area of 487,000 ha (1,202,890 ac) in northeastern Senegal. This reserve was created in 1972 to protect Sahelian fauna and to serve as a final destination for Sahelo-Saharan antelopes reintroduced in Senegal. Beginning as 440 ha (1,087 ac) in 2001, the Katané enclosure was given 200 more ha (494 ac) in 2009 (with fence materials and installation donated by the Exotic Wildlife Association in Texas) and expanded by another 600 ha (1,482 ac) in 2012, for a present size of 1,240 ha

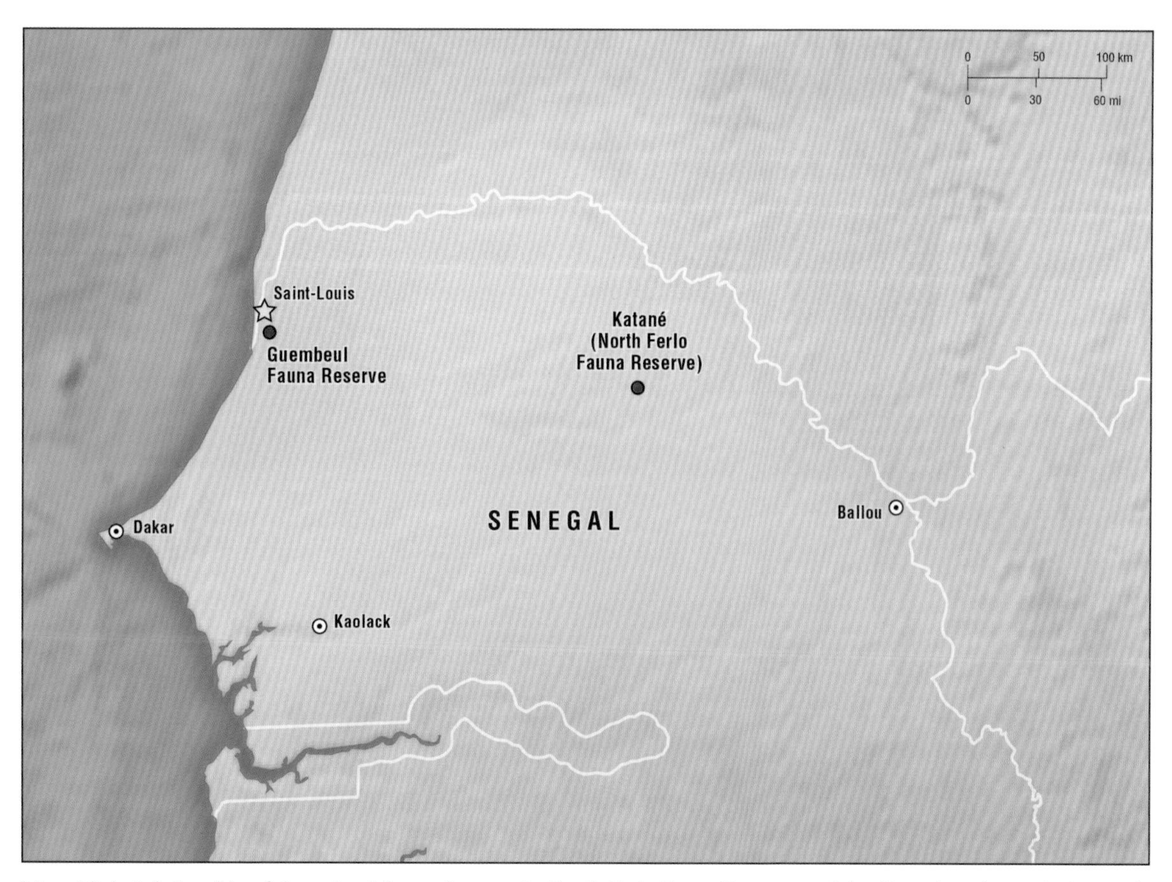

Map 18.4. Relationship of Guembeul Fauna Reserve to North Ferlo Fauna Reserve and the Katané enclosure in Senegal.

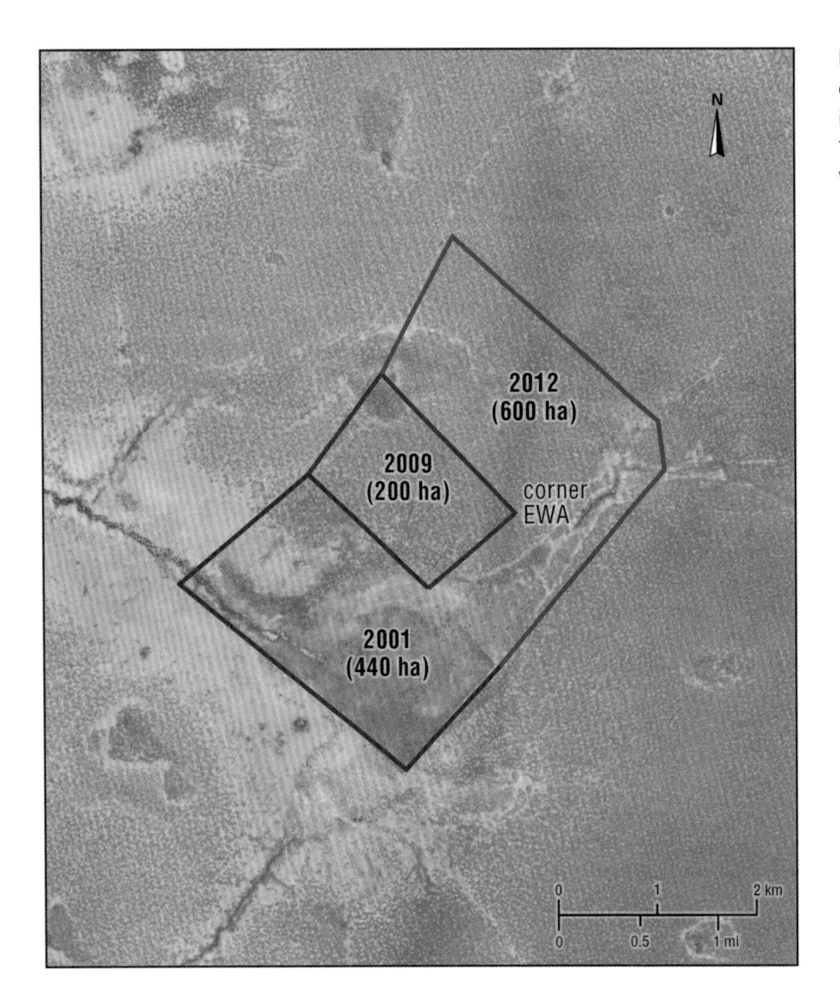

Map 18.5. Aerial view of Katané enclosure and its expansions, including the fencing sent from Texas, USA, to Senegal by the Exotic Wildlife Association (EWA).

(3,063 ac) (map 18.5). The Katané enclosure, in spite of its relatively small size, offers a variety of habitats and food resources to maintain nearly free-ranging populations of reintroduced Sahelo-Saharan antelope species (dorcas gazelle, mhorr gazelle, scimitar-horned oryx) as well as native red-fronted gazelle (*Eudorcas rufifrons*) and other native or reintroduced animals such as African spurred tortoise (*Centrochelys sulcata*) and warthog (*Phacochoerus aethiopicus*) and a variety of carnivores and birds.

Water is provided during the dry season (January–June). Woody species include a variety of trees and shrubs, most of which are more than 3 m (10 ft) tall (*Acacia ataxacantha*, *A. senegal*, *A. seyal*, *Adansonia digitata*, *Adenium obesum*, *Anogeissus leiocarpus*, *Balanites aegyptiaca*, *Boscia senegalensis*, *Calotropis procera*, *Combretum glutinosum*, *Commiphora africana*, *Crateva adansonii*, *Dalbergia melanoxylon*, *Grewia bicolor*, *Guiera senegalensis*, *Leptadenia hastata*, *Mitragyna inermis*, *Pterocarpus lucens*, and *Ziziphus mauritiana*). There is also an enormous variety of annual grasses and forbs: *Cassia tora*, *Cenchrus biflorus*, *Corchorus* spp., *Cucumis melo*, *Eragrostis* sp., *Pennisetum pedicellatum*, *Schoenefeldia gracilis*, and *Zornia glochidiata* being just a few of the kinds growing in the reserve (fig. 18.4). (See Abáigar et al. 2013a for a detailed description.)

The upward trend in numbers at Katané has been most encouraging. Although 4 females died in transport, the 3 that arrived survived and bred. From just 5 founders (2 males and 3 females, of which just 2 of the 3 females were adults) that arrived in 2003, the number grew to 7 in 2005, 13 in 2009, and 18 in 2013. The variety of habitats in the Katané enclosure let the animals move with the seasons depending on their requirements for food and shade. Mhorr gazelles are very shy, as shown by the way they avoid humans, but live in harmony with the other hoofed animals reintroduced at Katané, the scimitar-horned oryx and the dorcas gazelle (Abáigar et al. 2009).

Figure 18.4. Vegetative cover including grasses in the Katané enclosure at North Ferlo Fauna Reserve (photo by T. Abáigar/CSIC, courtesy of North Ferlo Fauna Reserve, Senegal).

There were also several other mhorr gazelles relocated within Senegal from Guembeul. Three went sometime before 1997 to the privately owned Bandia reserve (East 1999, cited in RZSS and IUCN Antelope Specialist Group 2014), and possibly others to the Fathala reserve, although neither reserve has this species now (RZSS and IUCN Antelope Specialist Group 2014).

Mauritania

The capture of a pair of mhorr gazelles at Guembeul in 2009 for relocation to a private facility in Mauritania was not actually a reintroduction project. The male died during transport, and the female lived alone at least until 2012.

R'Mila Royal Reserve

Two of the three reintroductions in Morocco have succeeded and one has failed (map 18.6). Most successful has been the reintroduction at R'Mila Royal Reserve. This reserve, covering 465 ha (1,149 ac)—of which 330 ha (815 ac) are occupied by gazelles—is 8 km (5 mi) north of Marrakech (map 18.7) and was created in 1982 (for details, see Cuzin et al. 2007).

Woody species include a variety of trees and shrubs such as *Acacia gummifera, Atriplex halimus, Eucalyptus torquata, Retama monosperma, Tamarix* sp., and *Ziziphus lotus* (fig. 18.5). Grass cover is dominated by *Stipa capensis.* For R'Mila, 6 gazelles (3 males and 3 females) from the EEP program (from Almería and from the Munich Zoo, where stock was derived from Almería) went in 1992. Between 1992 and 1996, additional mhorr gazelles were brought in from German zoos, and by 1997, the R'Mila total was up to 41 (Wiesner and Müller 1998, Wiesner and Müller, cited in RZSS and IUCN Antelope Specialist Group 2014).

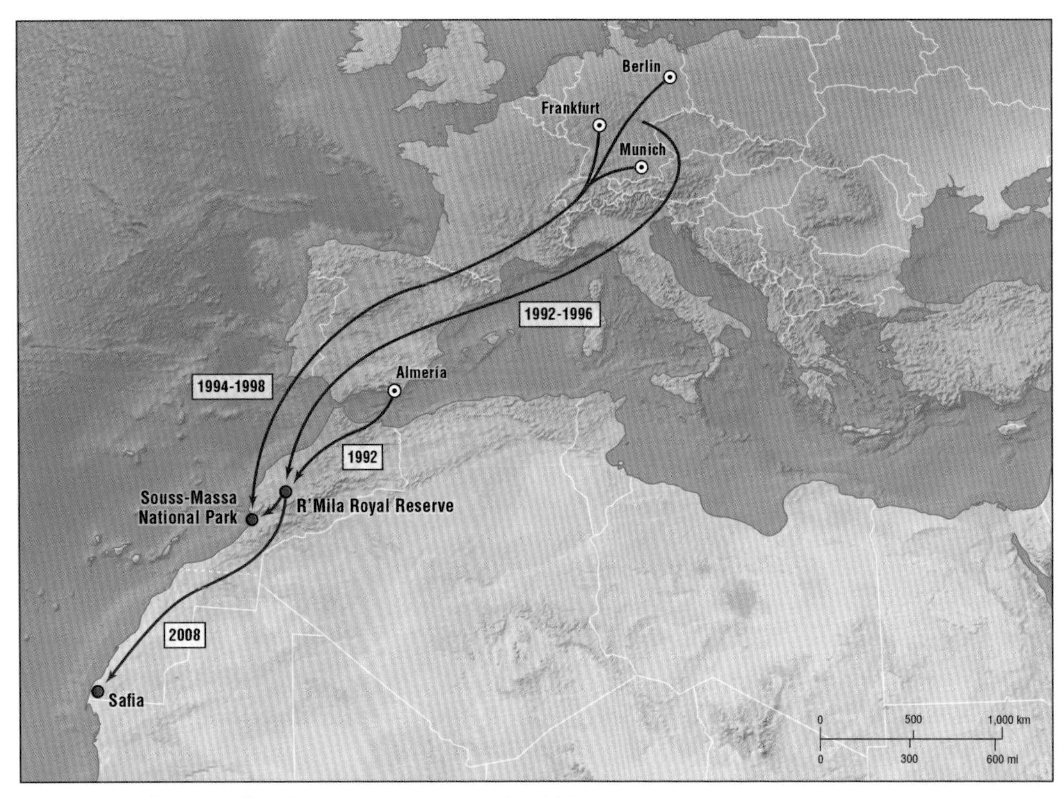

Map 18.6. Mhorr gazelle releases in Morocco, 1992–2008.

Map 18.7. Aerial view of R'Mila Royal Reserve showing its proximity to the city of Marrakech, Morocco.

Figure 18.5. Vegetative cover at R'Mila Royal Reserve with several of the mhorr gazelle males that it supports (photo by Mar Cano/CSIC, courtesy of R'Mila Royal Reserve, Morocco).

Figure 18.6. Dorcas gazelles pass a group of mhorr gazelles in R'Mila Royal Reserve (photo by Mar Cano/CSIC, courtesy of R'Mila Royal Reserve, Morocco).

By 2002, the number was 65 (Müller 2002), and by 2005 it was 110. By 2013, the R'Mila total had risen substantially, with reports of more than 158 (RZSS and IUCN Antelope Specialist Group 2014) in spite of the withdrawal of animals to stock two other reserves (Safia and Souss-Massa).

Besides the mhorr gazelles, a population of several hundred dorcas gazelles lives at the R'Mila reserve (fig. 18.6). Food and water are provided regularly, and there is constant supervision of the animals by rangers.

Rokkein Enclosure in Souss-Massa National Park

Between 1994 and 1998, 21 mhorr gazelles (13 males and 8 females) were transported from zoos in Berlin, Frankfurt, and Munich to the Rokkein reserve (2,000 ha, 4,940 ac) in Souss-Massa National Park (Wiesner and Müller 1998, Müller 2002) (map 18.8).

The vegetation in the reserve is dominated by reforestation of *Acacia cyanophylla*, *A. cyclops*, *Eucalyptus gomphocephala*, and native trees and shrubs like *Argania spinosa*, *Periploca laevigata*, *Retama monosperma*, and *Withania frutescens*. By 2002, the number of mhorr gazelles had decreased to 11 individuals (Müller 2002), and it remained almost stable until 2006, when 12 were reported (Cuzin et al. 2007). By 2013, there was not a mhorr gazelle in the Rokkein. Causes of their disappearance are unclear. At present, dorcas gazelles, addax (*Addax nasomaculatus*), and red-necked ostriches (*Struthio camelus camelus*) are present in the Rokkein enclosure, so there is no obvious reason why mhorr gazelles would not have survived, too (figs. 18.7 and 18.8).

Safia Enclosure in Safia Nature Reserve

In April 2008, 16 mhorr gazelles (7 males and 9 females) from the R'Mila Royal Reserve arrived at

Map 18.8. Aerial view of Souss-Massa National Park, Morocco, with its Rokkein reserve northern sector and its Arrouais reserve southern sector indicated.

Figure 18.7. Addax at Rokkein reserve in Souss-Massa National Park (photo by T. Abáigar/CSIC, courtesy of Souss-Massa National Park, Morocco).

Figure 18.8. Ostrich, including light-colored individuals, at Rokkein reserve in Souss-Massa National Park (photo by T. Abáigar/CSIC, courtesy of Souss-Massa National Park, Morocco).

Figure 18.9. Mhorr gazelle being released at the Safia enclosure, April 2008 (photo from CBD-Habitat Foundation, courtesy of Safia Nature Reserve, Morocco).

Figure 18.10. Large herd of mhorr gazelles in desert habitat typical of Safia Nature Reserve (photo by T. Abáigar/CSIC, courtesy of Safia Nature Reserve, Morocco).

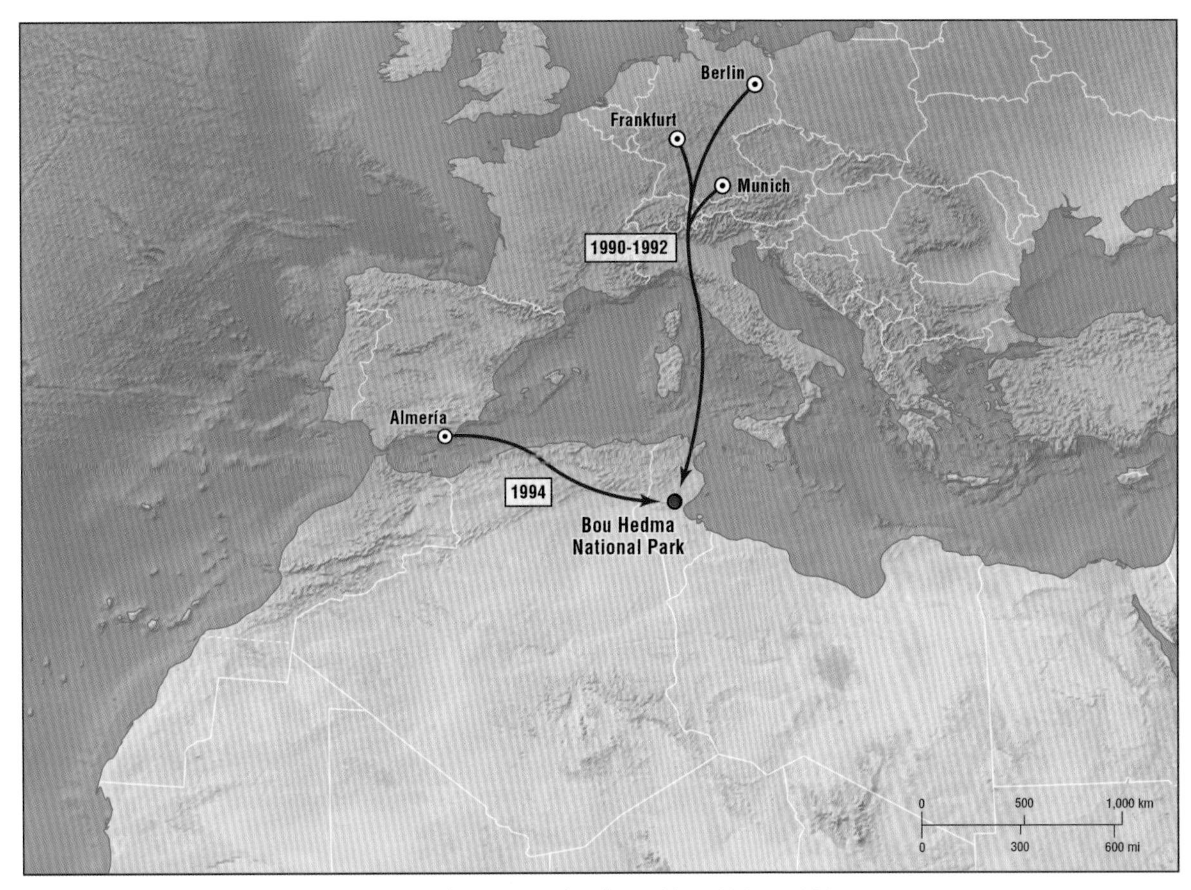

Map 18.9. Mhorr gazelle releases in Bou Hedma National Park, Tunisia, 1990 to 1994.

the Safia enclosure of the Safia Nature Reserve in far southern Morocco, close to the Mauritanian border (fig. 18.9; map 18.6). Three other gazelles died during transport.

The Safia enclosure was created in 2006 (with a size of 600 ha [1,482 ac] or somewhat more) and has typical desert vegetation (fig. 18.10). There are *oued* (seasonal watercourses, or wadis) and wet areas dominated by *Acacia raddiana*, which is the staple diet for mhorr gazelles there. Besides the mhorr gazelles, addax and red-necked ostrich were also reintroduced into the Safia enclosure. The reported totals for mhorr gazelle have continued to rise since the reintroduction. There were 20 in 2009, 26 the next year, 30 the next, and 41 in the enclosure by 2013 (data provided by CBD-Habitat).

Bou Hedma Biosphere Reserve and National Park

The releases at Bou Hedma Biosphere Reserve and National Park, Tunisia (map 18.9), began with the transport of 8 gazelles (2 males and 6 females) from German zoos (Berlin, Frankfurt, and Munich) in three transfers from 1990 to 1992 (Wiesner and Müller 1998, Jebali and Zahzah 2013).

Unfortunately, 4 (1 male and 3 females) died a few days after transport, so that by the end of 1993, the number was 4 (1 male and 3 females) (Jebali and Zahzah 2013) (fig. 18.11). Then in 1994, 14 mhorr gazelles (4 males and 10 females) arrived from Almería (Abáigar et al. 1997). This boosted the 1994 population to 21 individuals. After reaching a high of 28 animals in 2000, the population would have been expected to keep climbing. Instead, it varied for the next seven years (2001–2007). Then, it fell over the subsequent four years to its unsustainable 3 males (fig. 18.12). Jebali and Zahzah (2013), in a revision regarding the causes of mortality, explained the progression of this population by pointing first to predation by golden jackals (*Canis aureus*, 54 percent), then to injuries following accidents (27 percent), and finally to unknown causes (19 percent).

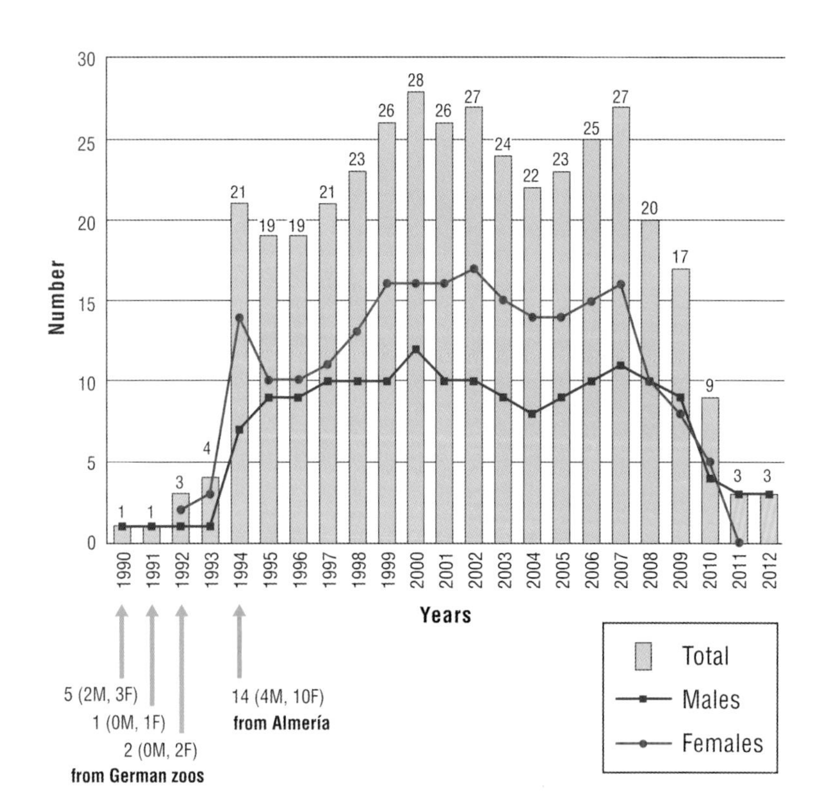

Figure 18.11. Rise and fall of mhorr gazelle population in Bou Hedma National Park, Tunisia (graph from Jebali and Zahzah 2013, modified to show releases).

Figure 18.12. The three remaining mhorr gazelles, all males, in the Bou Hedma Biosphere Reserve and National Park (photographed in the Haddej National Nature Reserve sector of the park, which used to have a connecting corridor to other Bou Hedma National Park habitat) (photo by Coke Smith, courtesy of Haddej National Nature Reserve, Tunisia).

Figure 18.13. Mhorr gazelle being released into open scrubland with mountains in the distance at Bou Hedma National Park, 1994 (photo by Mar Cano/CSIC, courtesy of Bou Hedma National Park, Tunisia).

Poaching is suspected for the remaining unidentified cases. By early 2017, there may have been only a single mhorr gazelle male left.

The Bou Hedma National Park area is open shrubland dominated by an extensive cover of *Acacia raddiana* (= *Acacia tortilis*) trees, which represent a vestige of pre-Saharan savanna similar to the Sahel (fig. 18.13). In the mountainous areas, *Juniperus phoenicea*, *Rhamnus lycioides*, *Rhus tripartita*, and *Rosmarinus officinalis* are found, while in the flat areas there are shrubs such as *Anabasis articulata*, *Artemisia herba-alba*, *Atriplex halimus*, *Lycium afrum*, *Pergularia tomentosa*, *Periploca laevigata*, *Retama raetam*, and *Rhanterium suaveolens*, plus the grass *Stipa tenacissima*. The mhorr gazelles were released in a total protected zone of 2,000 ha (4,940 ac). This protected area is fenced on three sides, while the mountain chain on the northwest side is, in principle, supposed to prevent escape of the gazelles because of the steep slopes. Jackals go in and out of the protected area through this mountainous part of the park. Reintroduced addax and scimitar-horned oryx, in addition to native dorcas gazelles, live with the mhorr gazelles.

How Can Introductions Succeed?

The dama gazelle is only one member of the greater fauna of the Sahara and its surrounding regions that has decreased dangerously in numbers and distribution (Durant et al. 2014). Like the addax, the dama gazelle seems to be following in the footsteps of the scimitar-horned oryx. This iconic oryx was declared "Extinct in the Wild" in 2000 (IUCN SSC Antelope Specialist Group 2014), as no wild representatives had been sighted since the 1980s (Newby 1988). Addax remain critically endangered, and central (*N. d. dama*) and eastern (*N. d. ruficollis*) subspecies of *Nanger dama* have only a few scattered remnants in the wild. In the case of the mhorr gazelle (*N. d. mhorr*), only reintroductions of captive-born animals will allow recovery of wild populations. It has been extinct in the wilds of Western Sahara since 1968 (Cano 1991). The wild Senegal representatives of dama gazelles, either the mhorr gazelle or the central dama gazelle, depending on where one considers the change between these two to have been (see chapter 2 on taxonomy and distribution), were last seen in the early 1970s (Poulet 1972, 1974). The first reintroduction of mhorr gazelles into the wild took place on May 22, 2015, when a group of mhorr gazelles was set free from the Safia enclosure in Safia Nature Reserve. As discussed in chapter 19, their progress is being monitored.

Until now, all the reintroductions of several species of Sahelo-Saharan antelopes, including gazelles—addax, scimitar-horned oryx, Cuvier's gazelle (*Gazella cuvieri*), dorcas gazelle, and mhorr gazelle—are at the stage where the animals are kept in fenced, protected areas of various sizes (several hundred to thousands of hectares) (Cano et al. 1993, Müller 2002, Woodfine et al. 2004, Jebali 2005, Gilbert and Woodfine 2008, Molcanova et al. 2011, Abáigar et al. 2009, 2013a), except for the single mhorr gazelle release mentioned, and none except the mhorr gazelle have been released into the wild within their native range. While this phase of keeping reintroduced species in (large) protected areas under human supervision is highly recommended at the earliest stage of a reintroduction project, keeping them this way as a final destination seems not to work all the time.

From the mhorr gazelle reintroductions described in this chapter, we could draw a conclusion relevant for the future of the conservation of this and other Sahelo-Saharan antelope species: as time passes with a species maintained in a fenced, protected area, risks arising from intrinsic (loss of genetic variability, sex ratio biases), extrinsic (food availability, predation, disease risk, human management), and random factors (accidents) are exacerbated, risking the positive development of the reintroduced population and its viability and survival. In spite of human intervention that could mitigate some of the negative effects of the above factors (adding new founders, providing supplemental food, eliminating predators, etc.), the risks remain and the development of a semicaptive population may not always be positive.

All the reintroduction projects described in this chapter started with a clear objective of releasing animals when conditions in their original areas of distribution were suitable. In practical terms, these conditions mean habitat availability, human social acceptance, and control of poaching. In spite of human impacts, suitable areas are, fortunately, still available for dama gazelles within their native distributional area. Moreover, the success of captive breeding programs ensures availability of animals for reintroductions (see the chapters about captive populations in section 3). So, human acceptance and poaching still remain the limiting factors that prevent the release of gazelles into the wild. In consequence, more efforts should be invested to ensure that there are suitable areas for reintroduction and to convince the people living in these areas that gazelles (as well as other animal species) are an essential part of their future.

Chapter 19

Release of Mhorr Gazelle Back to the Wild

Teresa Abáigar

At the time this book was in preparation, an important event for the preservation of the mhorr gazelle occurred. On May 22, 2015, came the long-awaited release of a group of mhorr gazelles into their original habitat. This was the first time ever that any of the captive-bred mhorr gazelles had been liberated back into the wild after their final disappearance as free-living animals in Western Sahara in 1968 (Cano 1991) (also see chapter 1 discussing the decline of the species and the loss of dama gazelles in Senegal in the early 1970s). At the time of this writing, the gazelles are spending their first days at liberty, and reports indicate that most of them are surviving. However, we believe it important to describe the first steps of this project because it marks a milestone toward recovery of the species.

This reintroduction project is the result of collaboration by a number of institutions, including the Haut Commissariat aux Eaux et Forêts et à la Lutte Contre la Désertification (HCEFLCD, Morocco); the Association Nature Initiative (ANI), a local non-governmental organization (NGO) based in Dakhla that was created in 2003 to preserve natural resources in the Oued Ed Dahab-Lagouira region (southern Morocco); the CBD-Habitat foundation, a Spanish NGO working with the HCEFLCD and the ANI on various projects to protect endangered species in the area; and the Estación Experimental de Zonas Áridas (EEZA–National Spanish Research Council), a center for research specializing in conservation and reintroduction of Sahelo-Saharan antelopes.

The decision to start this first experiment with the release of mhorr gazelles into the wild was made for several reasons: (1) the success of the mhorr gazelle reintroduction project in Morocco, with growing populations in both the R'Mila Royal Reserve and the enclosure at the Safia Nature Reserve; (2) adequate conditions in the Safia Nature Reserve regarding suitable habitat, reduced poaching (as a result of the patrolling program started in the area in 2008), and the lack of fixed human settlements (only temporary livestock) in the area; and (3) social acceptance of the project, with local associations working with authorities to preserve natural resources in the area. Moreover, this project is part of the Moroccan strategy for ungulate conservation (Cuzin et al. 2007).

Area of Release: Safia Nature Reserve

The Safia Nature Reserve (SNR), 3,075 km2 (1,230 mi2) in extent, is located south of the Oued Ed Dahab-Lagouira region in the Province of Aousserd (southern Morocco). The SNR is bordered to the south by Mauritania and to the west by the Atlantic Ocean. There are no permanent human settlements within the reserve, and the only human-related activity is seasonal livestock grazing by camels and goats. The single roadway maintained within the SNR is one to the west connecting Dakhla with the Mauritanian border.

The Safia Nature Reserve is characterized by slightly elevated relief with stone plains, chains of dunes, nebkhas (small sand dunes that form around vegetation), sebkhas (also spelled sebjas, depressions in the sand dunes), and wadis or *oueds* (dry riverbed channels) (fig. 19.1; map 19.1). The climate is arid Saharan, with sporadic rainfall of less than 100 mm per year (less than 4 in per year). Vegetation is typically Saharan. *Acacia raddiana* is the dominant tree in the wadis and small spring courses. It is the staple of the mhorr gazelle diet. There is also a variety of shrubs and annual plant species. Most important

Figure 19.1. A 50 ha enclosure with mhorr gazelles at Safia Nature Reserve. Note the various landforms inside and outside the enclosure (photo by T. Abáigar/CSIC, courtesy of Safia Nature Reserve, Morocco).

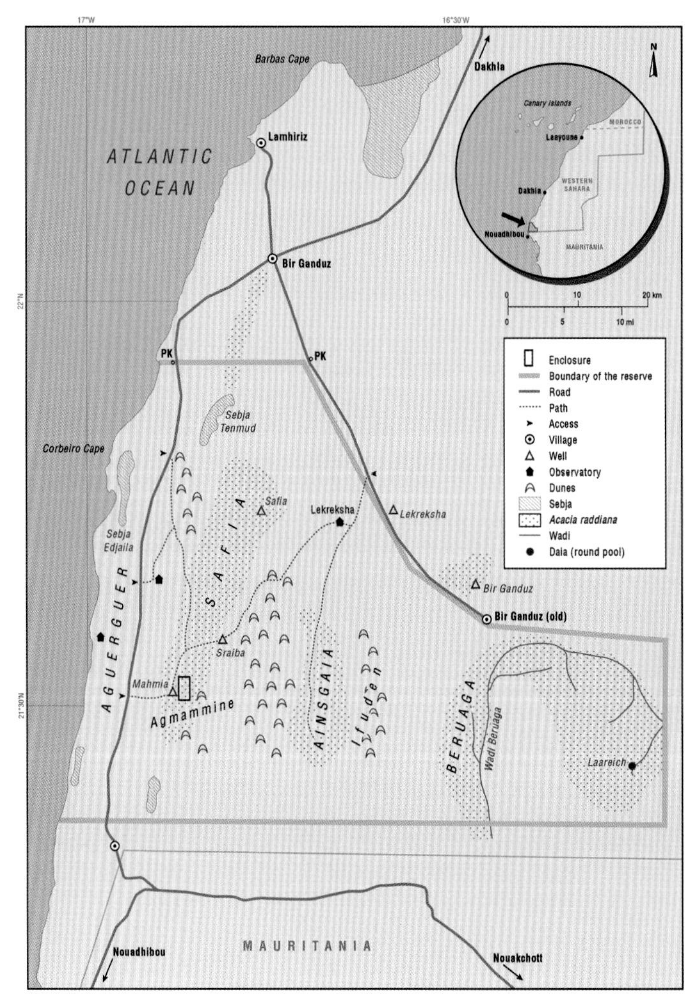

Map 19.1. Safia Nature Reserve showing the location of the fenced enclosures, the distribution of dunes, areas with favored *Acacia raddiana* browse, and features such as wadis, wells, and road (courtesy of T. Abáigar and the CBD-Habitat Foundation).

among these in terms of food for the gazelles are *Aizoon* sp., *Anabasis* sp., *Atriplex halimus*, *Capparis decidua*, *Launaea arborescens*, *Lycium intricatum*, *Maerua crassifolia*, *Nitraria retusa*, *Panicum turgidum*, *Salsola* spp., *Stipagrostis* spp., and *Ziziphus* sp.

The dorcas gazelle (*Gazella dorcas*) is the only ungulate species still free-ranging in the area after the complete extinction of the addax (*Addax nasomaculatus*), scimitar-horned oryx (*Oryx dammah*), and mhorr gazelle. Other important species that have disappeared are the red-necked ostrich (*Struthio camelus camelus*) and the cheetah (*Acinonyx jubatus*). Jackals (*Canis aureus*) and other small carnivores are still present.

Although hunting is completely forbidden, poaching is the main threat. However, regular patrolling established by CBD-Habitat and the ANI since 2008 has proven its effectiveness, with positive results for dorcas gazelle. This species is increasing in the patrolled area. In addition, an intensive awareness campaign is promoted regularly by the ANI, CBD-Habitat, and the HCEFLCD.

Moreover, in 2005, the HCEFLCD established within the SNR a 600 ha (1,482 ac) area protected by fencing in order to allow vegetation recovery before releasing, and keeping under semi-wild conditions, several native species that are extinct in the wild. The purpose of all this effort was to allow these species eventually to be reintroduced. Between 2008 and 2010, founder populations of mhorr gazelle from R'Mila Royal Reserve, Morocco, and of addax and red-necked ostrich from Souss-Massa National Park, Morocco, were stocked in the Safia enclosure (Bensouiba 2013). Before the May 2015 release of mhorr gazelles, the enclosure had a semi-wild population of 41 mhorr gazelles, 43 addax, and 21 red-necked ostriches.

Animals and Capture

Twenty-nine mhorr gazelles (16 males and 13 females) from the original group in the fenced area at Safia were isolated in a small enclosure of about 50 ha (124 ac) (fig. 19.2) at one end of the main area, which is now about 550 ha (1,358 ac). All the gazelles were adults (or at least more than 1 year old) except for three young males (two of 7 to 8 months old and one of 2 months old). The sex ratio of selected individuals was male biased, as it was in

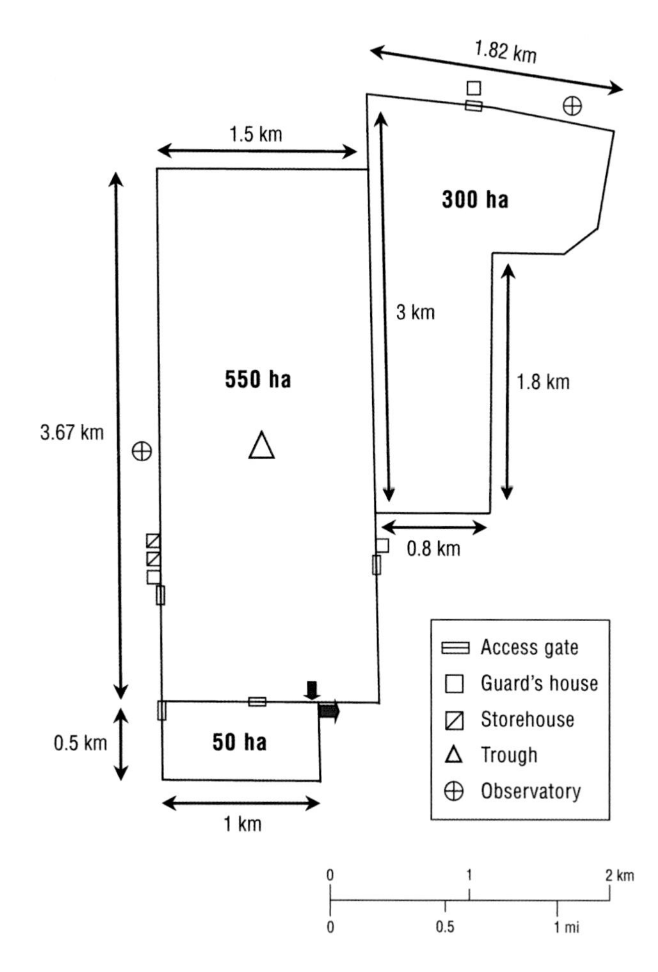

Figure 19.2. Enclosure arrangement at Safia Nature Reserve showing site of release (red arrow). Blue arrow shows where the gazelles used to go routinely between the 550 ha and the 50 ha space. The adjacent 300 ha was fenced in 2014 so that, in the future, it can be opened to the other 600 ha in order to give the animals more room.

the original group. The gazelles spent 11 days in the small enclosure before the final capture for release.

On May 17, the mhorr gazelles were handled using a drive corral built specifically for this purpose within the approximately 50 ha (124 ac) enclosure (fig. 19.3). In all, 25 of the 29 gazelles gathered for the release program were pushed through the series of progressively narrower spaces until they reached individual crates. Only 4 gazelles (2 males and 2 females), including the youngest juvenile male, were left in the 50 ha (124 ac) enclosure without manipulation. Next, each gazelle successfully put through the drive corral was separated and captured individually. The gazelles were caught manually by a keeper taking hold of the head at the base of the

Figure 19.3. Mhorr gazelles at the start of the drive corral specially built within the 50 ha enclosure for the release project at Safia Nature Reserve (photo by T. Abáigar/CSIC, courtesy of Safia Nature Reserve, Morocco).

Figure 19.4. This adult mhorr gazelle was caught manually and blindfolded in order to reduce stress (photo by T. Abáigar/CSIC, courtesy of Safia Nature Reserve, Morocco).

Figure 19.5. Collar, ear tag, and colored tape on horn are applied to a male mhorr gazelle during restraint (photo by M. Cedenilla, courtesy of CBD-Habitat Foundation).

horns and pulling the animal toward him. A mask was immediately put on the animal in order to reduce stress, and with the help of two other keepers, the animal was placed in the shade (fig. 19.4). Here, each gazelle was effectively immobilized by tying its legs with cords.

The capture started early in the morning and finished at 11:15 a.m., when the outside temperature had increased significantly. By that time, the gazelles' body temperatures had increased markedly and the risk of collapse was high.

Before rising temperatures halted the capture operation, 19 gazelles (8 males and 11 females) had been manually immobilized. For each of these gazelles, body temperature was measured, health condition examined, blood and feces sampled, and an anti-parasite drug administered. To aid identification after release, ears were tagged, colored plastic tape was placed on the horns (the left horn on males and the right horn on females), and photographs of individual traits (horns, white patch on the ventral neck, and pattern of the color on the back) were taken. In addition, 12 GPS-satellite collars were fitted (fig. 19.5), 5 on adult males and 7 on adult females, so that movements and activity of the gazelles could be monitored after release (fig. 19.6).

Four adult males were not immobilized because, as mentioned above, the capture stopped when outside temperatures increased. Therefore, these males continued through final release into the wild without addition of artificial identification marks. The 2 young males of 7 to 8 months old died before capture after a stampede against the fence of the drive corral caused skull fracture.

After capture, all of the 19 gazelles that had been immobilized, except 1 female (suffering from capture myopathy and returned to the main fenced area), were turned back to join the other gazelles in the approximately 50 ha (124 ac) enclosure and stayed there for a five-day settling period before complete release into the wild. During this time, 2 of the adult collared males died.

In summary, (a) 24 mhorr gazelles (12 males and 12 females) were released into the wild, of which 10 (3 males and 7 females) had GPS-satellite collars, 6 (3 males and 3 females) also had ear tags and horn tape but not collars, and 8 (6 males and 2 females) went without specific identification; (b) 2 (the two

Figure 19.6. Mhorr gazelle tagged, collared, and ready for release into the wild (photo by M. Cedenilla, courtesy of CBD-Habitat Foundation).

7- to 8-month-old males = 8 percent) out of the 25 gazelles put into the drive corrals died before capture, and another 2 (collared adult males = 8 percent) died after capture. For the 4 deaths, stress was the primary cause; either because of panic in the case of the 2 isolated young males, or because of stress suffered after manipulation in the case of the 2 collared adult males.

Release into the Wild

On May 22, 2015, a 48 m length of fence in a corner of the approximately 50 ha enclosure was removed so that the gazelles could go out (fig. 19.2). This corner was selected as the preferred site for the gazelles to exit because it was where the gazelles were used to passing from the main fenced area (about 550 ha, 1,358 ac) to the small enclosure (about 50 ha, 124 ac), as the two spaces were ordinarily open to each other. Also, this corner was where the gazelles spent most of their time after being confined to the 50 ha (124 ac) enclosure. The ground at the juncture between the fence and the outside was arranged so that the gazelles would not notice any difference that might inhibit exit.

The gazelles were not pushed to go out. During the night, they left on their own. By early the following morning, all the gazelles were out in the wild starting to explore their new surroundings.

The first week after release, while the animals still remained close to the fence of the familiar enclosure, a group of wild dogs attacked the gazelles, killing 7 individuals (6 males and 1 female). After this episode, the surviving group of 14 gazelles (5 males and 9 females) spent the next several months making exploratory movements in the reserve. Eventually, the gazelles selected two different areas at a distance of about 30 and 40 km (19 to 25 mi) north of the enclosure. The animals adapted well to the wild and to the environment, facing even persecution from jackals with success. They would travel together or in different groups separated from each other by a few kilometers.

Just when the monitoring results were very hopeful, a group of poachers entered the reserve at the beginning of October, provoking the gazelles to scatter. One group moved more than 130 km (81 mi) from the reserve to the Adrar Soutouff area. Another group moved to the vicinity of the border with Mauritania. The rest remained in the Safia reserve.

The field team is working on locating the different groups. Currently, three satellite collars are still operating of the ten deployed, which were of three different models. From the beginning, two collars never worked, probably because of a failure in the system of recharging batteries through solar panels. Of the other five collars that are not operative currently, one was deployed on a male that died during the early attack by wild dogs. Another collar was on a female that died two months after the release, and the other three stopped working during the dispersion caused by the poachers.

The Future

The Safia Nature Reserve (SNR) has a combination of features that make it a suitable place to conduct a first trial for returning the mhorr gazelle to the wild. The SNR has effective protection, no human settlements, and habitat suitable for the gazelles. The effective protection that has been established is particularly important, as are the public awareness campaigns that have reduced (although not eliminated) the main threat in the reserve (poaching). These efforts join with the favorable adaptation of mhorr gazelles to life at the SNR and with the obvious increase of reintroduced mhorr gazelles in the large fenced area. As a result, all the actors and organizations involved in the area's nature conservation are trying to give a chance to the species, the habitat, and the people to reverse the general tendency of biodiversity loss. May this release of mhorr gazelles back into the wild become the starting point for recovery of an entire ecosystem.

Although it is too early to predict the outcome of this reintroduction project, we think its success will be related to four important factors: (1) the success of the mhorr gazelles in asserting their natural ability to deal with the challenges that life in the wild imposes, including predators; (2) additional releases of mhorr gazelles in the same area in order to increase numbers as a way to enhance the medium- to long-term success of the project; (3) protection against harmful human activity (such as poaching or unregulated human incursion by livestock, dogs, garbage, wood exploitation, and tourism); and (4) a reevaluation of the importance of the area's natural resources and increased social awareness to promote sustainable development based on nature conservation. If these conditions can prevail, then mhorr gazelles can increase toward a bright future in the wild.

Project Execution and the Need for Many Dedicated Participants

A project like the one described for the mhorr gazelle would not be possible without the cooperation and participation of many people and institutions. These include the HCEFLCD: Zouhair Amhaouch, Abderrahim Essalhi, and their staff at Dakhla and Bir Ganduz, as well as guardians of the Safia enclosures (Hossein Jisso, Bilid Ajid, Ali Shmiss); CBD-Habitat and the Association Nature Initiative: Pablo Fernandez de Larrinoa, Cristina Martinez, Fernando Aparicio, Mohamed Lemine Samlali, Mohamed Lemine (coordinator), Sid Ahmed El Makki, Mohammed El Mokhtar Nafaa, Ali Lemdimigh, Mhammad Alifal, Taoufik El Balla, Hamady M'Bareck, Moulaye Haya, Abba M'Bareck; the Spanish Ministry of Agriculture, Fish and Environment: Luis Mariano Gonzalez and Francisco Garcia; the Parques Reunidos Foundation: Jesús Fernández; and Juan Lazcano.

In addition to the above named, the author extends sincere thanks to the participation of the regional government of Dakhla, to the government of the Province of Oued Ed Dahab-Lagouira, to the Province of Aousserd, to the authorities of Bir Ganduz, and to the Gendarmerie Royale for their support of the whole reintroduction process. This reintroduction action was developed under funding from the Man and the Biosphere Programme of UNESCO.

Chapter 20

Dama Gazelles in the Wild

John Newby, Tim Wacher, and Thomas Rabeil

Figure 20.1. Skull of a dama gazelle male found in the Termit region of Niger (photo by John Newby, copyright © John Newby/Sahara Conservation Fund).

The dama gazelle is one of the world's most threatened species (fig. 20.1). It is listed by the International Union for Conservation of Nature (IUCN) as Critically Endangered, and there are certainly fewer than 300 of these stunning animals left in the wild today. The Sahara Conservation Fund (SCF) has made this gazelle's survival one of its top priorities and is actively involved in its conservation. In November 2013, along with many others, the SCF was involved in a strategic planning workshop in Edinburgh to define a global recovery plan for the species based on a wide range of activities, including enhanced protection of wild populations, captive breeding, sperm banking, and public awareness initiatives.

Until fairly recently, the dama gazelle was a relatively widespread inhabitant of the vast, sparsely wooded grasslands of the Sahel region that borders

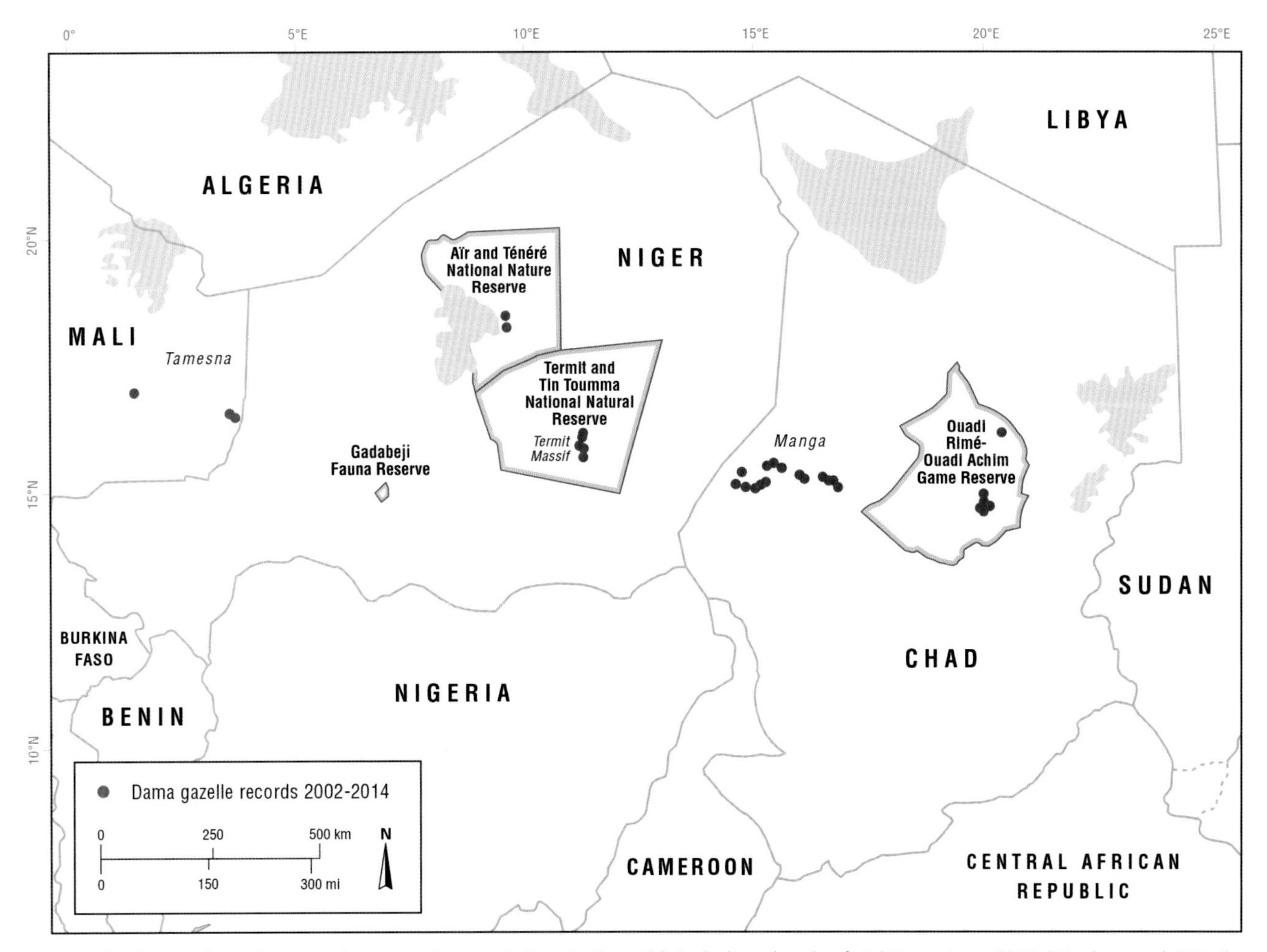

Map 20.1. Location of current dama gazelle populations in the wild, including details of sightings since 2001 (Wacher et al. 2014).

the great Sahara Desert of North Africa (Newby et al. 2016). However, as with the scimitar-horned oryx that shared this habitat, the arrival in the 1950s of fast, all-terrain vehicles and powerful weapons quickly wiped out the vulnerable plains-living animals. The relict populations have been largely pushed into marginal, rocky or subdesert habitats less frequented by people. It is in such areas, in Chad, Niger, and possibly Mali, that dama gazelles cling to survival, although even here they are increasingly threatened by poachers and habitat loss.

Today, truly wild dama gazelles can be found for certain in only four, tiny and widely dispersed populations in Chad and Niger (map 20.1). Survey work carried out by the SCF and its partners since 2001 has identified these four remaining strongholds: the Manga (fig. 20.2) and Ouadi Rimé–Ouadi Achim Game Reserve in Chad (figs. 20.3 and 20.4), and the Aïr Mountains (figs. 20.5 and 20.6) and the Termit Massif (figs. 20.7, 20.8, and 20.9) in Niger. In the early part of this century, a small number of dama gazelles could also be found in the Tamesna region straddling the border between Mali and Niger (fig. 20.10). However, the current situation there is unknown. In Sudan, where the gazelle was once quite common in the Wadi Howar region, the current situation is also unknown. In May 2015, 24 captive-bred dama gazelles were released in the Safia Nature Reserve in southern Morocco. To date, the results of this attempted reintroduction are inconclusive (see chapter 19).

Figure 20.2. An adult male dama gazelle leaving tracks through the sand in the Manga region of Chad (photo by Tim Wacher, copyright © Tim Wacher/ Zoological Society of London).

Figure 20.3. A pair of dama gazelles in Ouadi Rimé–Ouadi Achim Game Reserve grassland, Chad (photo by John Newby, copyright © John Newby/Sahara Conservation Fund).

Figure 20.4. Large group of dama gazelles in Ouadi Rimé–Ouadi Achim Game Reserve scrubland in 2012–2013 (photo by John Newby, copyright © John Newby/Sahara Conservation Fund).

Figure 20.5. Dama gazelle amid sparse vegetation in the Aïr Mountains, Niger (photo by Thomas Rabeil, copyright © Thomas Rabeil/Sahara Conservation Fund).

Figure 20.6. Aïr Mountains of Niger show as row upon row of ridges (photo by John Newby, copyright © John Newby/ Sahara Conservation Fund).

Figure 20.7. A camera trap in the Termit and Tin Toumma National Nature Reserve, Niger, caught this dama gazelle; its well-developed haunch stripe with a long vertical extension is characteristic of central and western members of this species (photo courtesy of Thomas Rabeil, copyright © Thomas Rabeil/ Sahara Conservation Fund).

Figure 20.8. Tufts of vegetation at Termit, Niger (photo by John Newby, copyright © John Newby/Sahara Conservation Fund).

Figure 20.9. Hills of rocks and sand, Termit, Niger (photo from John Newby, copyright © John Newby/Sahara Conservation Fund).

Figure 20.10. Dama gazelles were so rare in Tamesna, Mali, that this was the only chance during the 2005 survey flight for one quick picture, and a conference with the expedition scientific leader was called for verification (photo by Stéphane Bouju, copyright © Stéphane Bouju, Tamesna, Mali).

Whereas the gazelles in Ouadi Rimé, Aïr, and Termit are found in protected areas and as such benefit from a degree of security, the Manga, which lies between these sites, and the Tamesna to the far west, have no formal protection as yet. The rockier habitats of the Aïr Mountains and the Termit Massif —also home to aoudad (fig. 20.11)—appear to be atypical habitat for dama gazelles (fig. 20.12), but valued nonetheless as refuges from human persecution (RZSS and IUCN Antelope Specialist Group 2014).

Wildlife surveys carried out in the Manga region of western Chad illustrate the precarious state of the dama gazelle and the substantial logistics required to conserve and monitor it in the wild. Results from surveys in 2001 and 2010 demonstrated the presence of dama gazelles in the largest contiguous

Figure 20.11. Aoudad at Termit, Niger, on rocky slopes typical of their favored habitat (photo by Thomas Rabeil, copyright © Thomas Rabeil/Sahara Conservation Fund).

Figure 20.12. Young dama gazelle in atypical rocky habitat, Termit Massif, Niger (photo by Thomas Rabeil, copyright © Thomas Rabeil/Sahara Conservation Fund).

zone known, with up to 18 individuals encountered in 2001 and group sizes up to 8 strong seen in 2010 (Wacher and Newby 2010). On the basis of this and a further ground survey carried out in January and February 2014 (Wacher et al. 2014), a team of scientists from the SCF, the Zoological Society of London, and the Chadian wildlife department carried out an extensive aerial survey in February 2015 of the entire known range in the Manga region, with a simultaneous ground survey (Wacher et al. 2015). This survey located only two groups of dama gazelles, totaling four animals. Pasture in the target area was very sparse and mainly dry, with much of the more suitable wildlife habitat occupied by nomads and their livestock. Over the years, a large truck route linking Chad and Libya has also opened up, traversing the entire region. As a result of all this, at the time of the survey the known core dama habitat was in a marginal state. Those gazelles that were found were at the absolute northern edge of available dama habitat, squeezed by hard desert to the north, extensive livestock to the south, and heavy vehicle traffic nearby as a constant source of disturbance and possibly poaching.

While the aerial team was flying, the ground team found dama gazelle tracks east of former sightings in a greener area where Nubian bustard and dorcas gazelle were also noted in good numbers. No dama gazelles were seen, but with unsurveyed green pasture extending to the eastern horizon, there may well have been dama gazelles out there that had recently moved beyond the already very extensive survey zone.

Fieldwork on the dama gazelle has underlined the value of carrying out surveys at different times of the year. It also stresses the value of dialogue with local people and the information gleaned about the seasonal movements of pastoralists in the Manga. During the early part of the wet season, the herders move south to exploit fresh grazing, allowing the gazelles to occupy the zone the herders leave behind. As the wet season progresses and pastures extend farther north, the locals return to their traditional haunts and the dama gazelles move elsewhere.

Apart from the Manga, Chad's other main dama gazelle population, of about 50 animals, can be found in the center of the Ouadi Rimé–Ouadi Achim Game Reserve, a vast protected area roughly twice the size of Belgium and composed of thinly wooded grasslands and sandy, sub-desert habitats. Here, the dama gazelles occupy more typical habitat but are largely confined to the rugged dunes and wooded depressions by pressure from human disturbance and livestock. Unlike the smaller and more cosmopolitan dorcas gazelles, dama gazelles seem less able to cope with human beings and their activities, even when these are relatively benign. In recent times, efforts to conserve dama gazelles and their last remaining habitats have been boosted by the reintroduction of the scimitar-horned oryx and the accompanying monitoring, sensitization, and protection activities.

In Niger's Aïr Mountains, very little is known about the remaining, relict dama gazelle population because of more than three decades of rebellion and insecurity. In 2014, however, with the return of peace to northern Niger, the SCF and its partners from the wildlife department carried out a survey of the Takolokouzet Plateaus with the help of local guides and livestock herders. A total of nine dama gazelles, including a group of seven with an immature, and many fresh tracks were seen during two days of trekking in remote areas accessible only on foot. With this initial information in hand, further surveys will be carried out to assess the abundance and distribution of this most northerly of the known dama gazelle populations. A grid of camera traps will also be deployed and supervised by local guides.

Southwest of the Aïr, in the long but relatively low mountains of the Termit Massif, can be found Niger's other remaining population of dama gazelles. More than 10 years of systematic monitoring indicate a more or less stable population of some 30 to 40. This tiny population is nonetheless very vulnerable, confined as it is to certain undisturbed valleys and plateaus with very few possibilities of finding secure and viable habitat in surrounding areas. And although local conservation committees and community game guards have been put in place, lawlessness and the presence of arms and desert-going vehicles pose a constant threat.

Apart from the ever-present threat from poachers, the greatest challenge facing the dama gazelle and many of the Sahel's other species is habitat loss and competition from expanding livestock development (figs. 20.13–20.16). New wells and pumping stations are allowing access to hitherto lightly used pastures, causing an increase in livestock numbers (fig. 20.17).

Figure 20.13. Nomadic pastoralists and their livestock in Chad (photo by John Newby, copyright © John Newby/Sahara Conservation Fund).

Figure 20.14. Chadian horseman with a herd of goats (photo by John Newby, copyright © John Newby/Sahara Conservation Fund).

Figure 20.15. Chadian women traveling with donkeys (photo by John Newby, copyright © John Newby/Sahara Conservation Fund).

Figure 20.16. A camel load gets an adjustment (photo by John Newby, copyright © John Newby/Sahara Conservation Fund).

Figure 20.17. Camels crowd around for water (photo by John Newby, copyright © John Newby/Sahara Conservation Fund).

Figure 20.18. Dorcas gazelles (female *left*, male *right*) depend on much of the same habitat as the dama gazelles in Ouadi Rimé–Ouadi Achim Game Reserve, Chad (photo by John Newby, copyright © John Newby/Sahara Conservation Fund).

Figure 20.19. Scimitar-horned oryx adults in a Moroccan national park cluster around their young (photo by John Newby, copyright © John Newby/ Sahara Conservation Fund).

Figure 20.20. Female ostrich in North Africa (photo by John Newby, copyright © John Newby/Sahara Conservation Fund).

As a result, wildlife is pushed out into marginal habitat. On the relatively undisturbed land left to the gazelles, pasture is sparse, with shade and the water-rich plants, roots, and tubers needed by the gazelles to survive the long, waterless hot season being insufficient. A growing body of evidence suggests significant die-offs of both dama and dorcas gazelles (fig. 20.18) during this time of year. In the past, gazelles and larger animals, like the scimitar-horned oryx (fig. 20.19), addax, and ostrich (fig. 20.20), as well as all manner of smaller species, took refuge from the heat in densely vegetated wadis, where shade and water-rich plants were available. Most of these areas, some of which form natural "green walls" against desertification and encroachment of the Sahara, are now inhabited year round and in the worst cases are disappearing entirely under the axe for firewood or by clearance for seasonal agriculture (Rabeil et al. 2016).

Without better cooperation and understanding between wildlife and livestock interests, the outlook for Sahelo-Saharan wildlife is bleak. Overstocking of livestock and the associated degradation of pastures could also have catastrophic impacts on the pastoral economy. During severe droughts, livestock numbers will plummet, causing havoc among some of the world's poorest people. It is in everyone's interest that measures be taken to rationalize pastoral development and to seek a better balance with nature.

In the short to medium term, the best way to ensure that dama gazelles continue to survive and grow into more viable populations is through improved management of the protected areas in which the gazelles can still be found (map 20.1). In Niger, increased efforts to conserve both the Aïr and Ténéré National Nature Reserve and the Termit and Tin Toumma National Nature Reserve will have a positive impact. Similarly in Chad, improved management and protection of the Ouadi Rimé–Ouadi Achim Game Reserve is showing positive results. With projects in progress to protect both the Manga and Ennedi regions of Chad, and the Gadabeji region of Niger, opportunities also exist for reintroduction of dama gazelles. This would extend their range and, many hope, their viability.

The authors thank Bisbee's Fish and Wildlife Conservation Fund for its permission to expand here the article "Damas in Distress," which first appeared in its journal (Newby 2015). The authors also want to stress that it requires a collaborative effort by many organizations and people in order to continue the work mentioned here. Sincere gratitude is expressed to the following institutions for their significant contributions to ensuring the continued survival of the dama gazelle: the Wildlife Departments of both Chad and Niger; the IUCN Antelope Specialist Group; the Association of Zoos and Aquariums and the European Association of Zoos and Aquaria Ungulate Taxon Advisory Groups (Ungulate TAGs); the Addax and Oryx Foundation; the Zoological Society of London; the Royal Zoological Society of Scotland; the African Parks Network; the Smithsonian Conservation Biology Institute; the Estación Experimental de Zonas Áridas; the zoos of Al Ain (United Arab Emirates), Barcelona (Spain), Landau (Germany), Montpellier (France), Saint Louis (United States), and San Diego (United States); the Mohamed bin Zayed Species Conservation Fund; and the Private Department of His Highness Sheikh Mohamed bin Zayed Al Nahyan.

Conclusion

Elizabeth Cary Mungall

Discover the dama gazelles. This book delves into the past and present of this little-known complex of animals. In North Africa on the fringes of the Sahara Desert, from the darker, reddish mhorr gazelle near the Atlantic Coast, to the typical bright chestnut dama gazelle of the central Sahel, to the variously patterned roan addra or red-necked gazelle in Chad that reached into what became the Sudan, dama gazelles used to be a common element of the native fauna. Under a succession of local names, there used to be thousands of dama gazelles. As discussed in chapter 1 documenting decline, mhorr gazelles went extinct in Western Sahara in 1968. Their first release back into the wild came only in 2015. Scattered remnants farther east are estimated at only 300 animals or fewer (table 21.1). The situation is dangerously precarious—but not hopeless!

Overhunting and habitat loss as a result of human activity have created the present impasse. Now it is up to humans if the dama gazelles are to be saved. People who get informed and get involved can save pockets of habitat and build back the wildlife heritage of nations. People can do it if it becomes a priority in their range states. Thus, this book presents a variety of topics—from general behavior to genetics, from taxonomy to Texas homes, from removals from the wild to conservation in the wild, and more. May the book promote whatever interest might help people help this species.

The variety of topics discussed here is needed to support the variety of options still open to those who care about dama gazelles. Identifying and safeguarding what remains in the wild is certainly the best option, especially if the existing groups can become stable enough to allow increase. If wildlife authorities determine that reintroductions from captive populations outside Africa would be advantageous—to bolster the small wild numbers or to repopulate former habitat—then that can be done for the eastern race of dama gazelles. For them, there are numerous captive sources in zoos around the world, in parks, and especially on Texas ranches in the United States. As well as in zoos, mhorr gazelles are available even closer to the animals' home in a series of fenced West African breeding centers. They were established with the goal of one day reintroducing mhorr gazelles back to the wild. The central dama gazelle highlights the seriousness of today's plight. It has no captive herds to draw from outside Africa and no special breeding centers in Africa. There are just a very few wild wanderers and a very few that occasionally turn up in private ownership in Africa outside of any recorded conservation efforts.

Zoos and conservation-oriented parks keep the most closely managed groups of mhorr gazelles and other dama gazelles. These animals are as safe as confinement can make them. Health is monitored. Records of sires and dams are often kept in regional studbooks. Many of the matings are planned to minimize inbreeding. But one of the major problems for zoos is space. Often, only a few individuals can be accommodated in any single facility. Options for planned matings are limited, and only a portion of the natural behavior can be expressed.

Overcoming many of these problems is the system of private exotics ranches in the United States—especially in Texas. Even the smaller ranch populations are likely to have more room than zoos, and often more than parks as well. As pasture sizes increase, so does the range of natural behavior. The environment is more natural, and the animals forage for part, or even all, of their food. However, the greater freedom in a more natural environment comes at a cost, which includes various hazards in the wild. While the animals are usually secure from human threats, they have to deal to varying degrees with natural predators. In spite of the dangers, US populations of the eastern race of dama gazelle have flourished. A US census conducted by the Exotic Wildlife Association as of January 31, 2015, found 1,510 of these exotic dama gazelles, virtually all in Texas.

A comparison of the ranch situation to zoo

Table 21.1. Dama gazelle statistics at five sites where they remain in the wild compared with estimates from North African reintroduction and introduction sites

CHARACTERISTIC	PLACE (from west to east)					
	Northwest and west	Tamesna, Mali	Aïr-Ténéré, Niger	Termit, Niger	Manga, Chad	OROA, Chad
	Repatriated, fenced[a]	Wild	Wild	Wild	Wild	Wild
Estimated number of animals	231–235+	170[b, c, d]	At least 2[e]	50–60	At least 8	50–60[f]
Mean group size	?	3.33	1.00	4.13	2.14	2.6
Largest group seen	?	5	1	15	8	9
Survey or other year(s) as basis for estimate	2013	2005	2002	2007–2012	2001 and 2010	2011–2013
Area	4.65 to 24 km^2 (1.8 to 9.3 mi^2) for 5 sites in 3 countries	1,775 km^2 (685 mi^2) surveyed	77,360 km^2 (29,868 mi^2) surveyed in 1983–1984[e]	900 km^2 (347 mi^2) used by dama gazelles	4,000 to 7,000 km^2 (1,544 to 2,703 mi^2) used by dama gazelles	1,100 km^2 (425 mi^2) used by dama gazelles
Subspecies (according to Groves and Grubb 2011 boundaries)	Western dama gazelle (mhorr gazelle)	Central dama gazelle	Central dama gazelle	Central dama gazelle	Eastern dama gazelle (addra)	Eastern dama gazelle (addra)
SUBSPECIES TOTALS	Western dama gazelle (mhorr gazelle) 231–235+	Central dama gazelle 222–232			Eastern dama gazelle (addra) 58–68	
TOTAL NUMBER OF DAMA GAZELLES (estimated) on African rangeland 511–535+	FENCED Range of 3–158 for 5 sites 231–235+	WILD Range of 2–170 for 5 sites 280–300				

Note: OROA stands for Ouadi Rimé–Ouadi Achim Game Reserve in Chad. For further details on the populations in West African breeding centers, see chapter 18 on reintroductions.

[a]RZSS and IUCN Antelope Specialist Group 2014.

[b]Lamarque 2005.

[c]Lamarque 2006.

[d]Lamarque et al. 2007.

[e]Aïr-Ténéré survey in 1983–1984 estimated 150–250 dama gazelles (Grettenberger and Newby 1986).

[f]Down from 6,000 to 8,000 in the mid to late 1970s (RZSS and IUCN Antelope Specialist Group 2014, Thomassey and Newby 1990).

management shows that ranches as well as zoos make it a habit to found their herds from a variety of sources and to change out their breeding male every few years in order to reduce inbreeding. Unlike the situation in zoos, increasing ranch populations mean that it is increasingly difficult to know which matings provided which offspring. It is difficult to follow the lives of individual animals on ranches. When ranch populations enlarge into multi-male and multi-female groups, the animals make their own choices. If animals used to operating in as natural an environment as possible are wanted, these ranch animals would seem ideal. At least, they would seem ideal as long as they can adjust to close confinement and then new surroundings, if shipped and held in small settling enclosures before release. Fortunately, the success of a 2005 shipment from Texas ranches to the Middle East gives a promising sign for the future. Every one of the nearly four dozen addra, including fawns to adults, arrived in good condition.

All in all, Texas dama gazelle populations might seem to enjoy the ideal mix of safety and near-natural conditions. Ironically, their potential problem is politics. Any disruption to the captive-bred exemption they have under the US Endangered Species Act would result in owners debating whether to switch their holdings to other, less endangered, less regulated species. Permits allowing management are fine, but permits under this act have never provided the predictable avenue that ranchers can count on for the kind of long-range planning necessary to secure species for the future. Complete removal of these nonnative animals born and raised in the United States from listing under the US Endangered Species Act would encourage rancher activity with dama gazelles even more.

To help stabilize the situation for dama gazelles in the United States, concerned representatives from zoos, wildlife parks, and ranches met in 2015 to forge a special "Source Population Alliance." This cooperative initiative is intended to explore ways to benefit all types of participants and to create more opportunities for partners to coordinate their activities for dama gazelle conservation.

So, no single lifestyle for today's dama gazelles is ideal. A joint effort on all fronts is important. Range states and ranches, zoos and alliances are all called on to work together. The authors of this book with "boots on the ground" stress that the people living in the areas where dama gazelles still survive are a critical element in any long-term success. Their needs as well as their perceptions must be part of conservation plans.

While people work to rebuild sustainable, wild populations, they will also find advantages in encouraging the continuation of captive populations. The section of the book on zoos, parks, and ranches examines the backgrounds, aims, and potential of each of these environments for dama gazelles. This species needs help on a multi-faceted front in order to survive. That kind of combined effort can give dama gazelles the chance they need in order to make a comeback (fig. 21.1).

Figure 21.1. Mhorr gazelles in a West African breeding center awaiting the day when the Safia release efforts back into the wild can be expanded (photo courtesy of CBD-Habitat Foundation and T. Abáigar/CSIC, Safia Nature Reserve, Morocco).

Notes on Original Descriptions for Scientific Names

Compiled by Andrew C. Kitchener

La neuvième gazelle eſt un animal, qui, ſelon M. Adanſon, s'appelle *Nangueur* ou *Nanguer* au Sénégal, il a trois pieds & demi de longueur, deux pieds & demi de hauteur; il eſt de la forme & de la couleur du chevreuil, fauve ſur les parties ſupérieures du corps, blanc ſous le ventre & ſur les feſſes, avec une tache de cette même couleur ſous le cou; ſes cornes ſont permanentes comme celles des autres gazelles, & n'ont qu'environ ſix ou ſept pouces de longueur, elles ſont noires & rondes, mais ce qu'elles ont de très-particulier, c'eſt qu'elles ſont fort courbées à la pointe en avant, à peu près comme celles du chamois le ſont en arrière; ces nanguers ſont de très-jolis animaux & fort faciles à apprivoiſer; tous ces caractères, & principalement celui des petites cornes recourbées en avant, m'ont fait penſer

Current Taxonomy

Three subspecies:

Nanger dama dama (Pallas, 1766)

Nanger dama ruficollis (Hamilton Smith, 1827)

Nanger dama mhorr (Bennett, 1833)

Le Nanguer

Described by Buffon, 1764

Based on specimen collected by Michel Adanson from Senegal

Nangueur or Nanguer—Buffon

The ninth gazelle is an animal, which, according to M. Adanson, is called *Nangueur* or *Nanguer* in Senegal, it is three & a half feet in length, two & a half feet in height; it is the shape & color of the roe deer, tawny on the upper parts of the body, white on the belly & hindquarters, with a spot of the same color under the neck; its horns are permanent like those of other gazelles, & are about six or seven inches in length, they are black & round, but what they have in particular is that they are curved strongly forward at the tip, about like those of the chamois except that those of the chamois curve backward; these nanguers are very pretty animals & very easy to tame; all these characters, & chiefly that of the little horns curved forward, made me think of . . .

Antilope dama

Described by Peter Pallas, 1766

Based on Buffon, 1764

3. ANTILOPE *Dama* Plinii. Le Nanguer BUFFON. *hiſt. nat. Vol. XII. p.* 213. *t.* 34. Hujus ſolum caput cum cornibus vidi, e quo didici ſolam inter congeneres, dentes inferioris maxillæ tantum *ſenos* habere, quorum 2 medii latiſſimi & recta acie terminati ſunt, laterales parvi, ſublineares. Africana eſt, uti ſequentium pleræque, excepta ſexta, & forte quinta. Plurimæ tamen ex Africanis ſpeciebus Aſiæ quoque communes eſſe notandum.

Antilope dama Plinii—Pallas

3. ANTILOPE *Dama* of Pliny. Le Nanguer BUFFON. *hist. nat. Vol. XII. p.* 213. *t.* 34. I saw only the head with horns, from which I could learn only certain things, that it has only six teeth in the lower jaw, the two middle ones are very broad and end in straight lines, the lateral ones are small and sublinear. It is from Africa, as are many of the following [animals to be discussed next], except for the sixth, and perhaps the fifth species. Yet it should be noted that the majority of African species are common to Asia.

Antilope ruficollis

Described by Major Charles Hamilton Smith, 1927

Types in Senckenberg Museum, Frankfurt-am-Main

From Dongola, Sudan

The *Red-necked Antelope. (A. Ruficollis.)* There is so much similarity in the distribution of the colours, and even in the horns, between this species and *A. Dama*, that their possible identity requires an osculating location among the Antelopes. The species under consideration is about three feet high at the shoulder, of a light and elegant structure; the head rather broad across the orbits, tapers to a small mouth with ovine nostrils; the horns are near twelve inches long, with thirteen or sixteen small annuli, the superior third smooth; they are lyrated, with the tips turned inward and forward; beneath the eyes there is

Antilope mhorr

Described by Edward T. Bennett, 1833, in the *Proceedings of the Zoological Society of London*

Type is mounted skin of a male in the Natural History Museum, London (formerly alive at London Zoo)

From Wednun, near Tafilat, Mogador, Morocco

Expanded description by Bennett, 1835, in *Transactions of the Zoological Society of London*

TRANSACTIONS

OF

THE ZOOLOGICAL SOCIETY.

I. *On the* M'horr Antelope. *By* E. T. BENNETT, *Esq., F.L.S., Sec. Z.S.*

Communicated January 8, 1833.

PLINY appears, with one exception, to be the only author of antiquity who distinguishes the *Dama* of the classical ages by any tangible characters; and even his slight notices are confined to its transmarine origin[1], and the forward curvature of its horns[2]. In other writers the word, although of frequent occurrence, is accompanied only by vague epithets, indicative for the most part of gentleness, timidity, and velocity. Thus we have in Horace[3] the epithet "pavidæ"; in Virgil[4], "timidi"; in Martial, "molles[5], imbelles[6]"; in Seneca[7], "veloces"; and in Columella[8], "velocissimæ";—all applied to the *Damæ*, which appear, from the constant references made to them about that period, to have been well known at Rome in the times of the earlier Cæsars. The exception above noticed occurs in the fragment of the Halieuticon, generally ascribed to Ovid and at all events written by a contemporary author, and merely determines the animal to have had a fawn-coloured back, and to have been an object of the chase[9]. In this latter particular the writer, whoever he may have been, is confirmed by Virgil[10] and Columella[11]. It seems scarcely probable that an animal so well known, and com-

Gazelle dama permista

Described by Oscar Neumann, 1906

From Senegal, in Museum für Naturkunde, Berlin

Junior synonym of *N. d. dama*

Gazella dama permista nov. subsp.

Im allgemeinen ähnlich der *Gazella mhorr*[3]), aber das Weiß der Pygalgegend zieht sich jederseits als Spitze weit in den Körper hinein und läßt nur eine schmale, handbreite Verbindung zwischen dem Rotbraun des Rückensattels und dem der Keulen. Oberschenkel des Vorderbeins weiß. Erst von den Knien an ein hellerer brauner Streif auf der Vorderseite. Der rote Rückensattel reicht nicht so tief herab wie bei *mhorr*, sondern etwa bis zur Hälfte der Körperseite. Ganzer Nasenrücken weiß. Nur ein matter, dunkler Strich vom Auge nach vorn, aber kein scharfer schwarzer Fleck oder Fleckstrich. Nur wenig bräunliche Haare vorn zwischen den Hörnern. (Bei einem anderen Stücke ist fast der ganze Vorderkopf weiß, also auch kein dunkler Strich vor den Augen.)

Gazella dama permista nov. subsp.

In general similar to the *Gazella mhorr*[3]), but the white of the pygal region [rump] extends forward on each side to a point which is wide on the body and leaves only a narrow, hand breadth connection between the red-brown of the saddle-like color on the back and the legs. Upper portion of the forelegs white. Beginning from the knees, a lighter brown stripe runs down on the front side [of the forelegs]. The red saddle-like color on the back does not reach as low as in *mhorr*, but up to half of the side of the body. Whole of the nose is basically white. Only a weak, dark streak runs from the eye forward, but no sharp black spot or linear patch. Only a little brownish hair forward between the horns. (On another specimen the whole forehead is almost white, and also there is no dark line down from the eyes.)

Gazelle dama damergouensis

Described by Walter Rothschild, 1921

From Takoukout, Damergou, Niger River

Junior synonym of *N. d. dama*

3. **Gazella dama damergouensis** subsp. nov.

♂♀. Nearest to and intermediate between *G. d. permista* Neum. and *d. reducta* Heller., but distinguished from both by the rusty suffusion of all the white parts, and the greater extent of the rufous on legs. The horns are quite as thick as in *d. mhorr*, but longer in proportion.

1 ♂ ad., No. 118. Takoukout, Damergou, 1,550 ft., March 20, 1920. (Dead weight, 130 lb.; height at shoulder, $43\frac{1}{2}$ in. = 1,088 mm.; at rump, ditto; length of body, $37\frac{1}{2}$ in. = 938 mm.; girth of neck, $17\frac{1}{2}$ in. = 438 mm.; of body, $39\frac{3}{4}$ in. = 995 mm.) Right and left horns, 330 mm. = $13\frac{5}{16}$ in.

1 very old ♂ skull, procured at Damergou. Length of horns, 390 mm. = $15\frac{6}{16}$ in.

1 ♀ ad., No. 115. Takoukout, Damergou, 1,550 ft., March 7, 1920. (Height

Gazella dama lozanoi

Described by Eugenio Morales Agacino, 1934

From Cape Juby, Río de Oro, Western Sahara

Junior synonym of *N. d. mhorr*

7. **Gazella dama lozanoi** subsp. nov.

Tamaño grande; cuello con una característica mancha blanca en su tercio superior interno; dorso, hasta unos veinte centímetros del nacimiento de la cola, extremidades posteriores exteriormente y en su mayor parte, costados del cuerpo y cuello por todos sus lados, del color rojizo característico de las formas de *G. dama;* extremidades anteriores y mitad inferior de las posteriores, pecho, abdomen y cola en toda su extensión, asimismo como las nalgas y entrantes de la grupa, de color blanco; las proximidades de las pezuñas algo pardas, y éstas de color negro.

Gazella dama lozanoi subsp. nov.

Large size; neck with a characteristic white spot on the front upper third; back, until some twenty centimeters from the beginning of the tail, hind limbs externally and for the greater part, sides of the body and all around the neck, have the reddish color characteristic of the forms of *G. dama*; forelimbs and middle lower part of the back, chest, abdomen and all of the tail, likewise the same as the buttocks and forward from the croup, are white; the vicinity of the hoofs somewhat brown, and these black in color.

References

For further discussion and references, see chapter 2 on taxonomy and distribution.

Diagram of Dama Gazelle Physical Features

Elizabeth Cary Mungall

Hoofed animals have a special set of names for many parts of the body. Because some of these terms may be less familiar than others, a diagram is included here showing the main parts and color marks mentioned in the text. Modeling these is an adult male of the eastern dama gazelle.

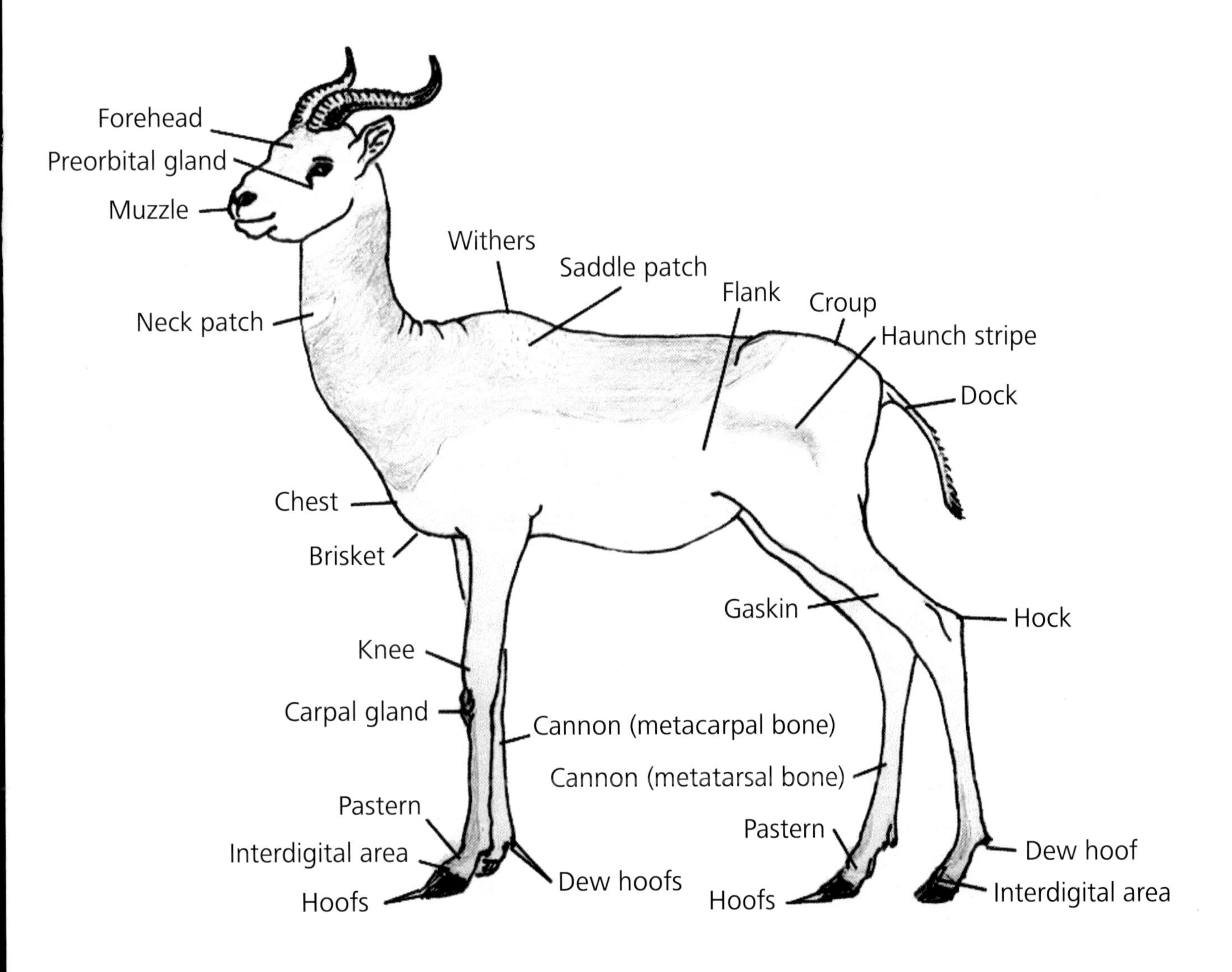

Glossary

This glossary defines words from the text that may be unfamiliar. Terms used in describing the conformation of hoofed animals are diagrammed in appendix 2.

Allele—a version of a gene. Some genes have numerous possibilities for a particular site on a strand of DNA.

Bout—a continuous, often short, session of a particular type of activity, such as nursing or grazing.

Cline—in genetics discussions, the gradual change in a trait, or the frequency of a trait, within a species over a geographic area.

Conspecific—another animal of the same species.

Control region—in genetics discussions, a region on a strand of DNA that has the ability to control the expression of linked genes.

Cytochrome b—a protein found in the mitochondria of cells that functions as part of the electron transport chain.

EAZA—acronym for the European Association of Zoos and Aquaria.

Erg—a large desert expanse of deep, rolling sand dunes.

Exotics—animals (or plants) living outside the native area of origin for the species. In Texas, the word "exotics" is usually taken to mean hoofed animals of foreign origin.

Fitness (genetic fitness)—from a biological perspective, fitness is the ability to survive to reproductive age and contribute fertile offspring to the next generation. Either animals or their genotypes can be discussed in terms of biological fitness.

Flehmen—German name for a gesture in which an animal lifts its nose, opens its mouth, and tests a scent. The upper lip is likely to be lifted more than it is normally. This is most common among males testing the estrous state of a female during courtship.

Forbs—non-woody, usually low-growing, broad-leaved plants, often regarded as weeds.

Founder animals—the animals that first started a population. In this sense, a population can be defined on any level; for example, the present population of all members of a species, or all members of a species living in a particular place (country, region, ranch, etc.).

Gene pool—the entire complement of genes carried by animals or plants alive in a population at a particular time.

Genetic bottleneck—a rapid, pronounced decrease in the size of a population that can cause a decrease in genetic diversity because of the loss of animals that happened to be carrying genes not carried by the surviving animals. It can be caused by a number of different factors including environmental events (such as floods, droughts, or disease) and human activities (such as over-hunting or habitat destruction).

Genetic drift—changes in the frequency of alleles in a population that are caused by chance. The smaller the population, the greater the impact genetic drift can have.

Genotype—the particular genetic makeup of an animal or plant; the sum total of the alleles that an individual happens to have.

Haplotype—a group of genes inherited from one parent.

Head-flagging—as performed by dama gazelles, a dominance display in which the gazelle stands erect with neck drawn back, raises its nose, and turns its head toward its target animal. The head may be turned away first in order to emphasize the turning back toward the target. The target animal is usually standing next to its partner, reciprocating with the same display, and often circling slowly.

Herd master See **master buck**.

Holocene—the present epoch of the Quaternary Period on the geologic time scale, extending from today back to the close of the Pleistocene Epoch approximately 10,000 years ago.

Holotype—the specimen on which an animal species or subspecies name is based.

Inbreeding—the mating of closely related individuals, such as fathers with daughters and siblings with each other, especially over multiple generations. Persistent inbreeding tends to increase the probability that an individual will inherit two copies of the same allele (form of a gene) for a particular trait, and therefore it increases the chance of expression of recessive, sometimes deleterious, traits.

Karyotype—the chromosomal number and arrangement carried by an individual.

Kernel Home Range (KHR)—a statistical method often used for estimating the area an animal uses, such as the animal's home range or core area. To estimate where the animal is likely to be at a given time, a utilization distribution is constructed using a normal distribution kernel density method.

Laufschlag—German name for a ritualized foreleg kick, as is often seen in the later stages of courtship. Often the leg is very stiff, but sometimes it is bent.

Lying out—The young of many hoofed species lie on the ground between maternal care sessions instead of constantly following the mother.

Master buck or **herd master**—a dominant adult male that controls an area and operates as a territorial male within this area. Often, a single male that has taken over an entire enclosure such as a zoo pen or a limited Texas pasture is called a master buck or herd master. In such cases, the types of behavior seen elsewhere between adult males at territorial borders may not be seen because the fences have become the borders.

Mitochondrial—having to do with cell structures called mitochondria, which are outside the cell nucleus and contain some genetic material and many enzymes important for cell metabolism (including enzymes responsible for conversion of food into usable energy). Mitochondria are passed down from mothers to their young through the female line only.

Mitochondrial DNA (mtDNA)—DNA embedded in cell structures called mitochondria that are outside the cell nucleus and are passed down only from mothers to their young, that is, through the female line only.

Mohor—in Arabic, *mohor* is the name for the chestnut color of many domestic horse foals. Because of the rich chestnut coat of the western subspecies of dama gazelle, "mohor" became its name. After Edward Bennett used a different spelling when giving this gazelle its taxonomic designation, this animal also became known as the mhorr gazelle.

Monotypic species—a species that is not divided into subspecies.

Natural markings—naturally occurring features that let an animal be recognized as an individual without incurring the risk of capturing it and marking it artificially with an ear tag or other device.

Neonate—a newborn animal.

OROA—acronym for the Ouadi Rimé–Ouadi Achim Game Reserve in central Chad.

Oued—seasonal watercourse, also called wadi.

Outbreeding—the random mating of unrelated individuals. It may produce more vigorous offspring (heterosis) because the same alleles for recessive, deleterious traits are less likely to be inherited from both parents.

Paratype—any specimen from a type series besides the **holotype**.

Phenotype—the expressed result of gene action on morphology and physiology.

Phyletic—of or relating to the evolutionary history of a particular group of organisms.

Phylogeographic—the study of the historical processes that may be responsible for the contemporary geographic distributions of individuals. Genetics, particularly population genetics, is the main method used in this kind of study.

Pleiotropic—refers to a single gene that produces multiple, and often seemingly quite different, phenotypic effects.

Polyphyletic—refers to a taxonomic group (such as a genus or species) to which more than one ancestral line has contributed. Therefore, more than one common ancestor can be part of the genetic makeup of members of the group.

Population structure—progressive genetic distinctiveness among populations.

Range states—countries that include native habitat for the species under discussion.

Reintroductions—human projects that return animals to native habitat.

Repatriations—human projects that put animals either into native habitat or into a region close to where their forebears originated.

Robertsonian translocation—a type of chromosome rearrangement that is formed by fusion of the long arms of two acrocentric chromosomes (chromosomes in which the centromere is near the end).

Sahel—the arid zone north and south of the Sahara. Moving away from the desert, it is first dry grassland, then savanna, then increasingly wooded. This band is wider in the south.

Sparring—low-intensity fighting. It does not look serious because the partners do not put much effort into it.

Stotting—a kind of jumping in which a hoofed animal springs upward with all four legs and then lands without intentionally going over anything.

Subadult—as defined in chapter 7 on growing up and growing old, and as used by ungulate expert Fritz R. Walther, the immature stage between adolescence and adulthood.

Swing-out movement—movement by which an animal prepares for a behavior to follow and which often goes in the opposite direction of the ensuing action. A swing-out movement may add momentum to the ensuing action.

Sympatric—occurring in the same geographic area, as for two different kinds of animals.

Ungulate—a hoofed mammal. This includes both even-hoofed animals like antelopes and deer and odd-hoofed animals like zebras and rhinoceroses.

Wadi—seasonal watercourse, also called *oued.*

References

Abáigar, T. 1993. Hematology and plasma chemistry values for captive dama gazelles (*Gazella dama mhorr*) and Cuvier's gazelles (*Gazella cuvieri*): Age, gender, and reproductive status differences. *Journal of Zoo and Wildlife Medicine* 24(2):177–84.

Abáigar, T., M. Cano, and C. Ensenyat. 2013a. Habitat preference of reintroduced dorcas gazelles (*Gazella dorcas neglecta*) in North Ferlo, Senegal. *Journal of Arid Environments* 97:176–81.

Abáigar, T., M. Cano, C. Ensenyat, M. Niaga, B. Youm, L. Kane, J. Gomis, T. Sarr, and H. Fernández. 2013b. Reintroduction project of dorcas gazelle in Senegal: Evaluation and assessment after eight working years. Abstract. Fourteenth meeting of the Sahelo-Saharan Interest Group, Agadir, Morocco, 2–5 May 2013.

Abáigar, T., M. Cano, G. Espeso, and J. Ortiz. 1997. Introduction of mhorr gazelle *Gazella dama mhorr* in Bou-Hedma National Park, Tunisia. *International Zoo Yearbook* 35:311–16.

Abáigar, T., M. Cano, A. R. Pickard, and W. V. Holt. 2001. Use of computer-assisted sperm motility assessment and multivariate pattern analysis to characterize ejaculate quality in Mohor gazelles (*Gazella dama mhorr*): Effects of body weight, electroejaculation technique and short-term semen storage. *Reproduction* 122(2):265–73.

Abáigar, T., M. Cano, and M. Sakkouhi. 2005. Evaluation of habitat use of a semi-captive population of Cuvier's gazelles (*Gazella cuvieri*) following release in Boukornine National Park, Tunisia. *Acta Theriologica* 50:405–15.

Abáigar, T., M. A. Domené, and F. Palomares. 2010. Effects of fecal age and seasonality on steroid hormone concentration as a reproductive parameter in field studies. *European Journal for Wildlife Research* 56(5):781–87.

Abáigar, T., and W. V. Holt. 2001. Towards the development of a genetic resource bank for the Mohor gazelle: Putting theory into practice. In *Cryobanking the Genetic Resource: Wildlife Conservation for the Future?*, edited by Paul Watson and William V. Holt. Taylor and Francis, New York. Pp. 123–39.

Abáigar, T., W. V. Holt, R. A. P. Harrison, and G. del Barrio. 1999. Sperm subpopulations in boar (*Sus scrofa*) and gazelle (*Gazella dama mhorr*) semen as revealed by pattern analysis of computer-assisted motility assessments. *Biology of Reproduction* 60(1):32–41.

Abáigar, T., J. Ortiz, M. Cano, C. Martinez-Carrasco, A. Albaladejo, and F. D. Alonso. 1995. Effect of mebendazole and ivermectin on the shedding of nematode eggs by three species of gazelles (*Gazella dama mhorr, G. cuvieri,* and *G. dorcas*). *Journal of Zoo and Wildlife Medicine* 26(3):392–95.

Abáigar, T., B. Youm, M. Niaga, C. Ensenyat, and M. Cano. 2009. The role of Senegal in the recovery of three Sahelo-Saharan antelope species: The case of the reintroduction of dorcas gazelle. *Gnusletter* 28:6–8.

Adanson, M. 1757. *Histoire naturelle du Sénégal.* Bauche, Paris. 275 pp.

Alados, C. L., and J. Escós. 1991. Phenotypic and genetic characteristics affecting lifetime reproductive success in female Cuvier's, dama and dorcas gazelles (*Gazella cuvieri, G. dama* and *G. dorcas*). *Journal of Zoology* 223:307–21.

———. 1992. The determinants of social-status and the effect of female rank on reproductive success in dama and Cuvier's gazelles. *Ethology Ecology & Evolution* 4(2):151–64.

Alba-Sánchez, F., S. Sabariego-Ruiz, C. Díaz de la Guardia, D. Nieto-Lugilde, and C. de Linares. 2010. Aerobiological behaviour of six anemophilous taxa in semi-arid environments of southern Europe (Almería, SE Spain). *Journal of Arid Environments* 74(11):1381–91.

Anonymous. 1996. *Exotic Hoofstock Survey.* Texas Agricultural Statistics Service and the Exotic Wildlife Association, Ingram. 4 pp.

Anonymous. 2010. Exotic Wildlife Association information sheet. Ingram, TX. 2 pp.

Anonymous. 2016. *Dama Gazelle Cross-Breeding Experiment—Update Report.* Al Ain Zoo in collaboration with the Royal Zoological Society Scotland, Al Ain, United Arab Emirates. 5 pp.

Araki, H., B. Cooper, and M. S. Blouin. 2007. Genetic effects of captive breeding cause a rapid, cumulative fitness decline in the wild. *Science* 318:100–103.

Arman, P., R. N. B. Kay, E. D. Goodall, and G. A. M. Sharman. 1974. The composition and yield of milk from captive red deer (*Cervus elaphus* L.). *Journal of Reproduction and Fertility* 37(1):67–84.

Arroyo Nombela, J. J., C. Rodriguez Murcia, T. Abáigar, and J. R. Vericad. 1990. GTG-banded karyotype of *Gazella dama mhorr* Bennett, 1833-cytogenetic relationship with other members of the subgenus *Nanger. Zeitschrift für Säugetierkunde* 55(3):194–201.

Audas, R. S. 1951. Game in northern Darfur. *Sudan Wild Life & Sport* 2:11–14.

AZA Antelope and Giraffe Advisory Group. 2008. *AZA Antelope and Giraffe Advisory Group Regional Collection Plan.* AZA, Silver Spring, MD. 170 pp.

Ballou, J. D., C. Lees, L. J. Faust, S. Long, C. Lynch, L. Bingaman Lackey, and T. J. Foose. 2010. Demographic and genetic management of captive populations. In *Wild Mammals in Captivity: Principles and Techniques for Zoo Management*, 2nd ed., edited by D. G. Kleiman, K. V. Thompson, and C. Kirk Baer. University of Chicago Press, Chicago. Pp. 219–52.

Bärmann, E. V., G. E. Rössner, and G. Wörheide. 2013. A revised phylogeny of Antilopini (Bovidae, Artiodactyla) using combined mitochondrial and nuclear genes. *Molecular Phylogenetics and Evolution* 67:484–93.

Barral, H. 1982. Le Ferlo des forages: Gestion ancienne et actuelle de l'espace pastoral. Office de la Recherche Scientifique et Technique Outre-Mer, Dakar, Senegal. 82 pp.

Benirschke, K. 1986. *Vanishing Animals.* Springer-Verlag, New York. 99 pp.

Bennett, E. T. 1833. Characters of a new species of antelope (*Antilope mhorr*) presented by E. W. A. Drummond Hay. *Proceedings of the Zoological Society of London* 1:1–2.

———. 1835. On the m'horr antelope. *Transactions of the Zoological Society of London* 1:1–8.

Ben Shaul, D. M. 1962. The composition of the milk of wild animals. *International Zoo Yearbook* 4:333–42.

Bensouiba, H. 2013. Acclimatation de la faune sahélo saharienne à la station de Safia. Thirteenth Annual Sahelo-Saharan Interest Group Meeting, 2–3 May 2013, Agadir, Morocco. Oral presentation.

Berlinguer, F., R. González, S. Succu, A. del Olmo, J. J. Garde, G. Espeso, M. Gomendio, S. Ledda, and E. R. S. Roldan. 2008. In vitro oocyte maturation, fertilization and culture after ovum pick-up in an endangered gazelle (*Gazella dama mhorr*). *Theriogenology* 69(3):349–59.

Blainville, H. de. 1816. Sur plusieurs espèces d'animaux mammifères, de l'ordre des ruminans. *Bulletin des Sciences par la Société Philomathique de Paris.* Pp. 73–82.

Boroviczeny, I. 1988. El rescate de las gacelas Mohor. *Boletín del Instituto de Estudios Almerienses.* Extra, pp. 43–50.

Borricand, P. 1945. Biogéographie des ongulés sahariens. *Notes Africaines* 27(July):9.

Bourbon, P. S. 1929. Résumé des recherches zoologiques de la mission Alger-Tchad. *Bulletin du Muséum National d'Histoire Naturelle* 2(1):283–91.

Bryden, H. A., ed. 1899. *Great and Small Game of Africa: An Account of the Distribution, Habits, and Natural History of the Sporting Mammals, with Personal Hunting Experiences.* Rowland Ward, London. 612 pp.

Buffon, G. L. Leclerc, Compte de, ed. 1764. *Histoire Naturelle, Générale et Particulière, avec la Description du Cabinet du Roi.* Vol. 12 (vol. 9 of quadrupeds), descriptions by le Compte de Buffon and L-J-M. Daubenton. De l'Imprimerie Royale, Paris. 452 pp.

Byers, O., C. Lees, J. Wilcken, and C. Schwitzer. 2013. The one plan approach: The philosophy and implementation of CBSG's approach to integrated species conservation planning. *WAZA Magazine* 14:2–5.

Cammaerts, D. 2003. Gazella dama mhorr: Observations éco-éthologiques en condition de semi-liberté d'un taxon éteint à l'état sauvage. Diplôme de DES en gestion de l'environnement, Belgium. 122 pp.

Cano, M. 1991. El antílope mohor (*Gazella* (*Nanger*) *dama mhorr* Bennett 1833) en cautividad. PhD diss., University of Granada, Granada, Spain. 760 pp.

———, T. Abáigar, and J. R. Vericad. 1993. Establishment of a group of dama gazelles (*Gazella (=Nanger) dama*) for reintroduction in Senegal. *International Zoo Yearbook* 32:98–107.

Cano Perez, M. 1984. Revision der Systematik von Gazella (*Nanger*) dama. *Zeitschrift des Kölner Zoo* 27:103–7.

———. 1988. Sobre las poblaciones de ungulados del Parque de Rescate de la Fauna Sahariana en el periodo 1971–1986. *Boletín del Instituto de Estudios Almerienses.* Extra, pp. 281–92.

———. 1990. El antilope mohor *Gazella* (=*Nanger*) *dama mhorr* Bennet, 1833 en cautividad. *Service Publications.* University of Granada, Granada, Spain.

Casado, A., R. de la Torre, E. López-Fernández, and B. Ruiz del Castillo. 1991. Hematologic and biochemical observations in *Gazella dama, Gazella dorcas* and *Gazella cuvieri. Comparative Biochemistry and Physiology Part B: Biochemistry* 99(3):637–40.

Cassinello, J. 2005. Inbreeding depression on reproductive performance and survival in captive gazelles of great conservation value. *Biological Conservation* 122:453–64.

Cassinello, J., T. Abáigar, M. Gomendio, and E. R. S. Roldan. 1998. Characteristics of the semen of three endangered species of gazelles (*Gazella dama mhorr, G. dorcas neglecta* and *G. cuvieri*). *Journal of Reproduction and Fertility* 113(1):35–45.

Cassinello, J., M. Gomendio, and E. R. S. Roldan. 2001. Relationship between coefficient of inbreeding and parasite burden in endangered gazelles. *Conservation Biology* 15(4):1171–74.

Cassinello, J., and I. Pieters. 2000. Multi-male captive groups of endangered dama gazelle: Social rank, aggression and enclosure effects. *Zoo Biology* 19(2):121–29.

Christie, M. R., M. L. Marine, R. A. French, and M. S. Blouin. 2012. Genetic adaptation to captivity can occur in a single generation. *Proceedings of the National Academy of Sciences of the United States of America* 109:238–42.

Cieslak, M., M. Reissmann, M. Hofreiter, and A. Ludwig. 2011. Colours of domestication. *Biological Reviews of the Cambridge Philosophical Society* 86(4):885–99.

Clark, B. 2001. Antelope programs in Senegal. In *Proceedings of the Second Annual Sahelo-Saharan Interest Group Meeting,* Almería, Spain, 9–10 May 2001, pp. 20–21.

———. 2002. *Oryx dammah* have achieved a second filial generation birth. In *Proceedings of the Third Annual Sahelo-Saharan Interest Group Meeting,* Bratislava, Slovakia, 20–22 May 2002, pp. 42–46.

Claro, F. 2004. Observations of antelopes in the Greater Termit Area, Niger, in 2002. *Antelope Survey Update* no. 9. IUCN/SSC Antelope Specialist Group Report. Pp. 47–51.

Cloudsley-Thompson, J. L. 1992. Wildlife massacre in Sudan. *Oryx* 26(4):202–4.

Cook, R. L. 1972. *Habitat Preference of Exotics.* Job no. 18, federal aid project no. W-76-R-15, job performance report. Texas Parks and Wildlife Department, Austin. 20 pp.

Corbet, G. B., and J. E. Hill. 1991. *A World List of Mammalian Species.* 3rd ed. British Museum (Natural History), London. 243 pp.

Crandall, L. S. 1964. *The Management of Wild Mammals in Captivity.* University of Chicago Press, Chicago. 769 pp.

Cretzschmar, P. J. 1826. Säugethiere. In *Altas zu der Reise im Nördlichen Afrika,* vol. 1, *Zoologie,* edited by Edward Rüppell. Senckenbergischen Naturforschenden Gesellschaft. Heinrich Ludwig Brönner, Frankfurt-am-Main, Germany. 78 pp.

Crnokrak, P., and D. Roff. 1999. Inbreeding depression in the wild. *Heredity* 83(pt. 3):260–70.

Cuzin, F. 2003. Les grands mammifères du Maroc méridional (Haut Atlas, Anti-Atlas et Sahara): Distribution, écologie et conservation. Doctoral thesis, Université de Montpellier II, Montpellier, France. 348 pp.

———, E. A. Sehhar, and T. Wacher. 2007. *Etude pour l'élaboration de lignes directrices et d'un plan d'action stratégique pour la conservation des ongulés au Maroc.* Vol. 1, *Rapport principal.* Haut Commissariat aux Eaux et Forêts et à la Lutte Contre la Désertification, Projet de Gestions des Aires Protégées, et Banque Mondiale, Global Environment Facility. 108 pp.

Darwin, C. 1876. *The Effects of Cross and Self-Fertilisation in the Vegetable Kingdom.* John Murray, London.

Devillers, P., J. Devillers-Tershuren, and R. C. Beudels-Jamar. 2006. *Gazella dama.* In *Sahelo-Saharan Antelopes Status and Perspectives: Report on the Conservation Status of the Six Sahelo-Saharan Antelopes,* 2nd ed., edited by R. C. Beudels, P. Devillers, R-M. Lafontaine, J. Devillers-Terschuren, and M.-O. Beudels. CMS SSA Concerted Action, CMS Technical Series Publication No. 11. UNEP/CMS Secretariat, Bonn, Germany. Pp. 57–70.

de Ybáñez, M. R. R., M. Goyena, T. Abáigar, M. M. Garijo, C. Martinez-Carrasco, G. Espeso, M. Cano, and J. M. Ortiz. 2004. Periparturient increase in faecal egg counts in a captive population of mohor gazelle (*Gazella dama mhorr*). *Veterinary Record* 154(2):49–52.

Dill, C. W., P. T. Tybor, T. McGill, and C. W. Ramsey. 1972. Gross composition and fatty acid constitution of blackbuck antelope (*Antilope cervicapra*) milk. *Canadian Journal of Zoology* 50(8):1127–29.

Dragesco-Joffé, A. 1993. *La vie sauvage au Sahara.* Delachaux and Niestlé, Paris. 240 pp.

Drake, N., and C. Bristow. 2006. Shorelines in the Sahara: Geomorphological evidence for an enhanced monsoon from paleolake Megachad. *The Holocene* 16(6):901–11.

Drüwa, P. 1985. Die Damagazelle (*Gazella dama* ssp. Pallas, 1767), einige Beiträge zur allgemeinen Biologie, Haltung and Zucht im Zoologischen Garten. *Der Zoologische Garten* (n.f.) 55:1–28.

Dubreuil, J. 1987. Coopération sahélienne dans le domaine de la conservation. In *Pour une gestion de la faune du Sahel,* edited by V. P. P. Vincke, G. Sournia, and E. Wangari. *Environnement Africain, Série Etudes et Recherches,* no. 120–21, MAB/ENDA/IUCN. Pp. 6–15.

Dupuy, A. R. 1967. Répartition actuelle des espèces menacées de l'Algérie. *Bulletin de la Société des Sciences Naturelles et Physiques du Maroc* 47(3–4): 355–84.

Durant, S. M., T. Wacher, S. Bashir, R. Woodroffe, P. De Ornellas, C. Ransom, J. Newby, T. Abáigar, M. Abdelgadir, H. El Alqamy, J. Baillie, M. Beddiaf, F. Belbachir, A. Belbachir-Bazi, N. E. Bemadjim, A. A. Berbash, R. Beudels-Jamar, L. Boitani, C. Breitenmoser, M. Cano, P. Chardonnet, B. Collen, W. A. Cornforth, F. Cuzin, P. Gerngross, G. Haddane, M. Hadjeloum, A. Jacobson, A. Jebali, F. Lamarque, D. Mallon, K. Minkowski, S. Monfort, B. Ndoassal, B. Niagate, G. Purchase, S. Samaïla, A. K. Samna, C. Sillero-Zubiri, A. E. Soultan, M. R. Stanley Price, and N. Pettorelli. 2014. Fiddling in biodiversity hotspots while deserts burn? Collapse of the Sahara's megafauna. In *Diversity and Distributions* 20(1):114–22, edited by D. Richardson. doi: 10.1111/ddi.12157.

East, R. 1990. *Antelopes: Global Survey and Regional Action Plans, Part 3. West and Central Africa.* IUCN–World Conservation Union, Gland, Switzerland. 173 pp.

EAZA (European Association of Zoos and Aquaria). 2004. http://www.eaza.net.

Effron, M., M. H. Bogart, A. T. Kumamoto, and K. Benirschke. 1976. Chromosome studies in the mammalian subfamily Antilopinae. *Genetica* 46:419–44.

Einarsen, A. S. 1948. *The Pronghorn Antelope and Its Management.* Wildlife Management Institute, Washington, DC. 238 pp.

Emanoil, M., and IUCN. 1994. *Encyclopaedia of Endangered Species.* Gale Research, Detroit. 1230 pp.

Espeso, G. 2015. SPARKS dataset for the *Nanger dama mhorr* international studbook. Unpublished data.

Espeso Pajares, G. 2008. Dama gazelle EEP Annual Report 2006. In *EAZA Yearbook 2006*, edited by D. de Man, W. van Lint, K. Garn, and B. Hiddinga. EAZA Executive Office, Amsterdam, Netherlands. Pp. 719–20.

Estes, P. D. 1991. *The Behavior Guide to African Mammals: Including Hoofed Mammals, Carnivores, Primates.* University of California Press, Berkeley. 611 pp.

Faust, L. J., and S. D. Thompson. 2000. Birth sex ratio in captive mammals: Patterns, biases, and the implications for management and conservation. *Zoo Biology* 19:11–25.

Flower, S. S. 1931. Contributions to our knowledge of the duration of life in vertebrate animals. *Proceedings of the Zoological Society of London* 5:145–234.

———. 1932. Notes on recent mammals of Egypt with a list of the species recorded from that kingdom. *Proceedings of the Zoological Society of London* 1:369–450.

Frankham, R. 1995a. Conservation genetics. *Annual Review of Genetics* 29:305–27.

———. 1995b. Inbreeding and extinction: A threshold effect. *Conservation Biology* 9:792–99.

———. 2003. Genetics and conservation biology. *Comptes Rendus Biologies* 326:S22–S29.

———. 2008. Genetic adaptation to captivity in species conservation programs. *Molecular Ecology* 17:325–33.

———. 2010. Inbreeding in the wild really does matter. *Heredity* 104:124.

Frankham, R., J. D. Ballou, and D. A. Briscoe. 2002. *Introduction to Conservation Genetics.* Cambridge University Press, Cambridge. 617 pp.

Garde, J. J., A. J. Soler, J. Cassinello, C. Crespo, A. F. Malo, G. Espeso, A. Gomendio, and E. R. S. Roldan. 2003. Sperm cryopreservation in three species of endangered gazelles (*Gazella cuvieri, G. dama mhorr,* and *G. dorcas neglecta*). *Biology of Reproduction* 69(2):602–11.

Gilbert, T., and T. Woodfine. 2008. *The Reintroduction of the Scimitar-Horned Oryx* Oryx dammah *to Dghoumes National Park.* Tunisia report to members of the European Endangered Species Programme for scimitar-horned oryx. 7 pp.

Gomendio, M., J. Cassinello, and E. R. S. Roldan. 2000. A comparative study of ejaculate traits in three endangered ungulates with different levels of inbreeding: Fluctuating asymmetry as an indicator of reproductive and genetic stress. *Proceedings of the Royal Society B–Biological Sciences* 267:875–82.

Gonzalez, R., F. Berlinguer, G. Espeso, F. Ariu, A. del Olmo, J. J. Garde, M. Gomendio, S. Ledda, and E. R. S. Roldan. 2008. Use of a neuroleptic in assisted reproduction of the critically endangered mohor gazelle (*Gazella dama mhorr*). *Theriogenology* 70(6):909–22.

Grandin, T., and C. Johnson. 2006. *Animals in Translation: Using the Mysteries of Autism to Decode Animal Behavior.* Harcourt, New York. 359 pp.

Gray, J. E. 1846. On the arrangement of the hollow-horned ruminants. *Annals and Magazine of Natural History Including Zoology, Botany and Geology* 18:227–33.

Grettenberger, J. F., and J. E. Newby. 1986. The status and ecology of the dama gazelle in the Aïr and Ténéré National Nature Reserve, Niger. *Biological Conservation* 38:207–16.

Groves, C., and P. Grubb. 2011. *Ungulate Taxonomy.* Johns Hopkins University Press, Baltimore. 336 pp.

Hamilton Smith, C. 1827. Supplement to the Order Ruminantia. In *The Animal Kingdom Arranged in Conformity with Its Organization, by the Baron Cuvier, Member of the Institute of France, etc., with Additional Descriptions of All the Species Hitherto Named, and of Many More Not Before Noticed,* vol. 4, *The Class Mammalia Arranged by the Baron Cuvier, with Specific Descriptions*, edited by E. Griffith, C. Hamilton Smith, and E. Pidgeon. G. B. Whittaker, London. Pp. 33–428.

Harmel, D. E. 1980. *Statewide Census of Exotic Big Game Animals*. Job no. 21, federal aid project no. W-109-R-3, job performance report. Texas Parks and Wildlife Department, Austin. 33 pp.

Hassanin, A., F. Delsuc, A. Ropiquet, C. Hammer, B. Jansen van Vuuren, C. Matthee, M. Ruiz-Garcia, F. C. Catzeflis, V. Areskoug, T. T. Nguyen, and A. Couloux. 2012. Pattern and timing of diversification of Cetartiodactyl (Mammalia, Laurasiatheria), as revealed by a comprehensive analysis of mitochondrial genomes. *Comptes Rendus Biologies* 335(1):32–50. doi:10.1016/j.crvi.2011.11.002.

Heim de Balsac, H. 1958. La faune en région prédésertique dans le Nord de l'Afrique: Facteurs de dégradation, moyens de sauvegarde. In "Préservation de la faune sauvage en région semi-aride." Special issue, *La Terre et la Vie*. Compterendu du symposium de Caracas de l'IUCN. 105(4):301–4.

Heringa, A. C. 1990. Mali. Chap. 4 in *Antelopes: Global Survey and Regional Action Plans, Part 3, West and Central Africa*, edited by R. East. IUCN, Gland, Switzerland. Pp. 8–14.

Heringa, A. C., U. Belemsobgo, C. A. Spinage, and G. W. Frame. 1990. Burkina Faso. Chap. 14 in *Antelopes: Global Survey and Regional Action Plans, Part 3, West and Central Africa*, edited by R. East. IUCN, Gland, Switzerland. Pp. 61–68.

Holt, W. V., T. Abáigar, and H. N. Jabbour. 1996. Oestrous synchronization, semen preservation and artificial insemination in the Mohor gazelle (*Gazella dama mhorr*) for the establishment of a genome resource bank programme. *Reproduction Fertility and Development* 8(8):1215–22.

Ibáñez, B., E. Moreno, and A. Barbosa. 2011. No inbreeding effects on body size in two captive endangered gazelles. *Zeitschrift für Säugetierkunde* 76:748–54.

———. 2013. Parity, but not inbreeding, affects juvenile mortality in two captive endangered gazelles. *Animal Conservation* 16:108–17.

In Tanoust. 1930. *La chasse dans le pays saharien et sahélien de l'Afrique occidentale français et de l'Afrique équatoriale française*. Comité Algérie-Tunisie Maroc, Paris. 208 pp.

ISIS (International Species Information System). 2015. Accessed September 1, 2015. https://www.species360.org.

IUCN (International Commission for Zoological Nomenclature). 1929. Opinion 108. Suspension of rules for Gazella 1816. *Smithsonian Miscellaneous Collections* 73(6):15.

———. 2002. *Technical Guidelines on the Management of Ex-Situ Populations for Conservation*. Approved at the 14th meeting of the programme committee of council. Gland, Switzerland.

———. 2012. *IUCN Guidelines for Reintroductions and Other Conservation Translocations*. IUCN/SSC, Gland, Switzerland. 16 pp.

IUCN SSC Antelope Specialist Group 2014. *Oryx dammah*. The IUCN Red List of Threatened Species, 2008, version 2014.1. Accessed July 16, 2014. http://www.iucnredlist.org.

Javier Cuervo, J., M. Dhaoui, and G. Espeso. 2011. Fluctuating asymmetry and blood parameters in three endangered gazelle species. *Mammalian Biology* 76(4):498–505.

Jebali, A. 2003. Rapport sur la réintroduction de l'oryx algazelle (*Oryx dammah*) et de la gazelle dama (*Gazella dama mhorr*) dans la Réserve de Faune du Ferlo Nord (RFFN), Sénégal. Mission avril 2003. Muséum National d'Histoire Naturelle de Paris/MAVA. 32 pp.

———. 2005. The re-introduction of scimitar-horned oryx, *Oryx dammah*, and dama gazelle, *Gazella dama mhorr* to Ferlo, Senegal: Two years after. In *Proceedings of the Sixth Annual Sahelo-Saharan Interest Group Meeting*, Haute Touche, France, 11–13 May 2005, pp. 84–90.

———. 2008. Déclin de la faune sahélo-saharienne et tentative de réintroduction d'antilopes dans des habitats restaurés: Cas de l'oryx algazelle (*Oryx dammah*) et de la gazelle dama (*Gazella dama mhorr*) dans la Réserve de Faune du Ferlo Nord (Sénégal). Doctoral thesis, Muséum National d'Histoire Naturelle de Paris, Paris. 394 pp.

Jebali, A., and K. Zahzah. 2013. *Causes et trajectoires de la mortalité chez deux gazelles réintroduites en Tunisie: La gazelle mohor* (Nanger dama mhorr) *au Parc National de Bou Hedma et la gazelle de l'Atlas* (Gazelle cuvieri) *dans le Parc National de Bou Kornine*. Rapport pour la Dir. Général des Forets, Tunis, Tunisia. 28 pp.

Jenness, R., and R. E. Sloan. 1970. The composition of milks of various species: A review. *Dairy Science Abstracts* 32(10):599–612.

Johnson, L., and E. C. Mungall. 2016. Historic shipment to Arabia. *Exotic Wildlife* (Spring–Summer), 21–23.

Joleaud, L. 1929. Etudes de géographie zoologique sur la Berbérie: Les ruminants, V, Les gazelles. *Bulletin de la Société Zoologique de France* 54:438–57.

Kacem, S. B. H., H-P. Müller, and H. Wiesner. 1994a. *Gestión de la faune sauvage et des parcs nationaux en Tunisie*. Imprimeries Réunies de Tunisie, Tunis, Tunisia. 305 pp.

———. 1994b. *Gestión de la faune sauvage et des parcs nationaux en Tunisie: Réintroduction, gestion et aménagement.* Deutsche Gesellschaft für Technische Zusammenarbeit, Eschborn, Germany. 305 pp.

Kephart, S. R. 2004. Inbreeding and reintroduction: Progeny success in rare *Silene* populations of varied density. *Conservation Genetics* 5:49–61.

Kitts, W. D., I. M. Cowan, J. Bandy, and A. J. Wood. 1956. The immediate post-natal growth in the Columbian black-tailed deer in relation to the composition of the milk of the doe. *Journal of Wildlife Management* 20(2):212–14.

Kriska, M. A. 2001. *Contribution à l'inventaire chorologique des biogéocènoses de l'Aïr et du Tamesna nigérien.* Memoires et travaux de l'Ecole Pratique des Hautes Études, no. 24. EPHE, Montpellier, France. 132 pp.

Lamarque, F. 2005. *Détermination du statut de conservation des gazelles dama dans le Sud Tamesna.* Rapport de mission en République du Mali, 6–18 February 2005. Programme 2004–2 du projet ASS-CMS/FFEM and Saint Louis Zoo. Office National de la Chasse et de la Faune Sauvage. 70 pp.

———. 2006. *Dénombrement aérien des gazelles dama dans le Sud Tamesna.* Rapport de mission en République du Mali, 14–24 November 2005. Programme 2004–2 du projet ASS-CMS/FFEM and Saint Louis Zoo. Office National de la Chasse et de la Faune Sauvage.

Lamarque, F., and B. Niagaté. 2004. A few data on the Sahelo-Saharan gazelles in the Tamesna region, Mali. *Antelope Survey Update* no. 9. IUCN/SSC Antelope Specialist Group Report. Pp. 24–25.

Lamarque, F., A. A. Sid'Ahmed, S. Bouhu, G. Coulibaly, and D. Maïga. 2007. Confirmation of the survival of the critically endangered dama gazelle (*Gazella dama*) in south Tamesna, Mali. *Oryx* 41(1):109–12.

Larrasoaña, J. C., A. P. Roberts, and E. J. Rohling. 2013. Dynamics of Green Sahara Periods and their role in hominin evolution. *PLoS ONE* 8(10):e76514. doi:10.1371/journal.pone.0076514.

Lataste, F. 1885. Étude de la faune des vertébrés de Barbarie. *Actes de la Société Linnéenne de Bordeaux* 39:129–299.

Latta, R. G. 2008. Conservation genetics as applied evolution: From genetic pattern to evolutionary process. *Evolutionary Applications* 1:84–94.

Lavauden, L. 1926. *Les vertébrés du Sahara: Eléments de zoologie saharienne.* Guinard, Tunis, Tunisia. 200 pp.

Laycock, G. 1966. *The Alien Animals.* Natural History Press, Garden City, NY. 240 pp.

Lhote, H. 1946. Observations sur la répartition actuelle et les mœurs de quelques grands mammifères du pays Touareg. *Mammalia* 10(1):25–56.

Linnaeus, C. 1758. *Systema naturae per regna tria naturae, secundum classis, ordines, genera, species cum characteribus, differentiis, synonymis, locis.* 10th ed. Vol. 1. Laurentii Salvii, Stockholm. 824 pp.

Loggers, C. O., M. Thévenot, and S. Aulagnier. 1992. Status and distribution of Moroccan wild ungulates. *Biological Conservation* 59:9–18.

Luisa Silva, T., R. Godinho, D. Castro, T. Abáigar, J. Carlos Brito, and P. Celio Alves. 2015. Genetic identification of endangered North African ungulates using noninvasive sampling. *Molecular Ecology Resources* 15:652–61.

Malbrant, R. 1952. *Faune du Centre Africain Français (mammifères et oiseaux).* 2nd ed. Paul Lechevalier, Paris.

Manlius, N. 1996. Biogéographie et écologie historique de quelques grands mammifères terrestres et sauvages en Egypte, depuis le Pléistocène final jusqu'à nos jours. Doctoral thesis, Muséum National d'Histoire Naturelle, Paris. 364 pp.

Maydon, H. C. 1923. North Kordofan to south Dongola. *Geographical Journal* 61:34–41.

Mayr, E., and P. D. Ashlock. 1991. *Principles of Systematic Zoology.* 2nd ed. McGraw-Hill, New York.

Mbouyou Boulende, T. 2011. Étude de la dynamique et du comportement alimentaire des gazelles *Gazella dama mhorr* en semi-liberté dans la Réserve Spéciale de Faune de Guembeul. Mémoire de diplôme de Master II, Université Cheikh Anta Diop, Dakar, Sénégal. 30 pp.

Molcanova, R., T. Wacher, K. Zahzah, and A. R. Fekih. 2011. Scimitar-horned oryx population management (PN Sidi Toui) and addax monitoring (PN Djebil and Senghar). *Gnusletter* 29(2):27–31.

Montgomery, M. E., L. M. Woodworth, P. R. England, D. A. Briscoe, and R. Frankham. 2010. Widespread selective sweeps affecting microsatellites in *Drosophila* populations adapting to captivity: Implications for captive breeding programs. *Biological Conservation* 143:1842–49.

Morales Agacino, E. 1934. Mamíferos colectados por la expedición L. Lozano en el Sahara español. *Boletín de la Sociedad Española de Historia Natural* 34:449–56.

Moreno, E., M. B. Ibáñez, and A. Barbosa. 2011. Mother traits and offspring sex in two threatened gazelle species in captivity. *Journal for Nature Conservation* 19(3):148–53.

Moreno, E., A. Sane, J. Benzal, B. Ibáñez, J. Sanz-Zuasti, and G. Espeso. 2012. Changes in habitat structure may explain decrease in reintroduced mohor gazelle population in the Guembeul Fauna Reserve, Senegal. *Animals* 2:347–60.

Müller, H. P. 2002. Overview of the situation of Sahelo-Saharan antelope in Morocco. In *Proceedings of the Third Annual Sahelo-Saharan Interest Group Meeting,* Zamocka, Slovakia, 20–22 May 2002.

Mungall, E. C. 1978. *The Indian Blackbuck Antelope: A Texas View.* Kleberg Studies in Natural Resources. 184 pp.
———. 1980. Courtship and mating behavior of the dama gazelle (*Gazelle dama,* Pallas 1766). *Zoologische Garten* 50(1):1–14.
———. 2004a. Dama gazelle phone census. *Exotic Wildlife* (January–February). P. 11.
———. 2004b. *Submission for the Comment Period on Proposed Listing of Scimitar-Horned Oryx, Addax, and Dama Gazelle under the Endangered Species Act.* Exotic Wildlife Association, Ingram, TX. 27 pp.
———. 2007a. *Exotic Animal Field Guide: Nonnative Mammals in the United States.* Texas A&M University Press, College Station. 265 pp.
———. 2007b. Ranch research significant for zoo success of dama gazelles. In *Proceedings of the 36th National Conference of the American Association of Zoo Keepers.* AAZK, Galveston, TX. Pp. 81–87.
———. 2010a. Free-ranging exotic hoofed stock trends in Texas: 1966–present. Abstract. In *Invasive Species: The Next Great Threat to Wildlife and Habitat.* Texas Chapter of the Wildlife Society, Galveston. Pp. 77–78.
———. 2010b. Miracle milk—results of the latest Second Ark Foundation research project. *Exotic Wildlife* (July–August). P. 35.
———. 2010c. Variations on a theme. *Exotic Wildlife* (November–December). P. 26.
———. 2013. WOW! Look at those home range patterns. *Exotic Wildlife* (Fall). Pp. 9, 11.
———. 2015. Dama collaring. *Bisbee's Conservation Journal* 2:78–85.
Mungall, E. C., K. Kyle, and S. A. Smith. 2007. Management implications of social maturation among captive dama gazelles. Abstract. In *Human-Wildlife Interactions.* Texas Chapter of the Wildlife Society, Beaumont. P. 60.
Mungall, E. C., and W. J. Sheffield. 1994. *Exotics on the Range: The Texas Example.* Texas A&M University Press, College Station. 265 pp.
Neumann, O. 1906. Über einige Gazellen und Kuh-Antilopen. *Sitzungsberichte der Gesellschaft Naturforschender Freunde zu Berlin.* Pp. 237–47.
Newby, J. 1974. *The Ecological Resources of the Ouadi Rimé–Ouadi Achim Faunal Reserve, Chad.* Report to UNDP/FAO Wildlife Conservation and Management Project CHD/69/004.
———. 1980. Action plan for Sahelo-Saharan fauna of Africa. Report to IUCN/WWF. In *WWF Yearbook 1980–81,* 466–74.
———. 1981. Is this the last chance for North Africa's fauna? World Wildlife Fund Monthly Report, June, project 1624. In *Antelopes Sahelo-Saharan,* 135–42.
———. 1982. *Avant-projet de classement d'une aire protégée dans l'Aïr et le Ténéré (République du Niger).* Rapport pour l'IUCN/WWF, October. 123 pp.
———. 1988. Aridland wildlife in decline: The case of the scimitar-horned oryx. In *Conservation and Biology of Desert Antelopes,* edited by A. Dixon and D. Jones. Christopher Helm, London. Pp. 146–66.
———. 2015. Damas in distress. *Bisbee's Conservation Journal* 1:58–67.
Newby, J., T. Wacher., S. M. Durant, N. Pettorelli, and T. Gilbert. 2016. Desert antelopes on the brink: How resilient is the Sahelo-Saharan ecosystem? In *Antelope Conservation in the 21st Century: From Diagnosis to Action.* Conservation Science and Practice Series. Wiley-Blackwell, Oxford, United Kingdom. Pp. 253–79.
Ortiz, J., R. R. de Ybanez, T. Abáigar, M. Goyena, M. Garijo, G. Espeso, and M. Cano. 2006. Output of gastrointestinal nematode eggs in the feces of captive gazelles (*Gazella dama mhorr, Gazella cuvieri* and *Gazella dorcas neglecta*) in a semiarid region of southeastern Spain. *Journal of Zoo and Wildlife Medicine* 37(3):249–54.
Osborn, D. J., and I. Helmy. 1980. *The Contemporary Land Mammals of Egypt (including Sinai).* Fieldiana Zoology new series, no. 5. 579 pp.
Osborn, D. J., and J. Osbornová. 1998. *The Mammals of Ancient Egypt.* Vol. 4, *The Natural History of Egypt.* Aris & Phillips, Warminster, United Kingdom. 213 pp.
Pallas, P. S. 1766. *Miscellanea zoologica quibus novae imprimis atque obscurae animalium species describuntur et observationibus iconibusque illustrantur.* Van Cleef, The Hague, Netherlands.
Pervinquière, L. 1912. *La tripolitaine interdite.* Ghadames Hachette, Paris. 251 pp.
Petric, A. 2012. *AZA North American Regional Studbook for Addra Gazelle* (Nanger dama ruficollis). Saint Louis Zoo, Saint Louis, MO. 223 pp.
Petric, A., and E. Spevak. 2012. *Addra Gazelle* Nanger dama ruficollis *AZA Species Survival Plan, Population Analysis & Breeding and Transfer Plan.* Saint Louis Zoo, Saint Louis, MO.
Pickard, A. R., T. Abáigar, D. I. Green, W. V. Holt, and M. Cano. 2001. Hormonal characterization of the reproductive cycle and pregnancy in the female Mohor gazelle (*Gazella dama mhorr*). *Reproduction* 122:571–80.
Pickard, A. R., W. V. Holta, D. I. Green, M. Cano, and T. Abáigar. 2003. Endocrine correlates of sexual behavior in the Mohor gazelle (*Gazella dama mhorr*). *Hormones and Behavior* 44(4):303–10.
Poilecot, P. 1996a. La faune de la Réserve Naturelle Nationale de l'Aïr et du Ténéré: L'autruche. In *La Réserve Naturelle de l'Aïr et du Ténéré (Niger),* edited by F. Giazzi. IUCN, Gland, Switzerland. Pp. 181–264.

———. 1996b. La faune de la Réserve Naturelle Nationale de l'Aïr et du Ténéré: La gazelle dama (ou biche Robert)—*Gazella dama dama* (Pallas, 1766). In *La Réserve Naturelle de l'Aïr et du Ténéré (Niger)*, edited by F. Giazzi. IUCN, Gland, Switzerland. Pp. 195–201.

Poulet, A. R. 1972. Recherches écologiques sur une savane sahélienne du Ferlo septentrional, Sénégal: Les mammifères. *La Terre et la Vie* 26:440–72.

———. 1974. Recherches écologiques sur une savane sahélienne du Ferlo septentrional, Sénégal: Quelques effets de la sécheresse sur le peuplement mammalien. *La Terre et la Vie* 1:124–30.

Rabeil, T., J. Newby, M. H. Hacha, and A. A. Moustapha. 2016. *Rapport de mission dans la Réserve de Faune de Ouadi Rimé–Ouadi Achim.* Mission report, Sahara Conservation Fund, Chad. 18 pp.

Rabeil, T., T. Wacher, J. Newby, A. Harouna, and A. Matchano. 2013. *Monitoring Dama Gazelles* Nanger dama *in the Termit Massif (Niger).* Project progress report part I.

Ralls, K., and J. Ballou. 1983. Extinction: Lessons from zoos. In *Genetics and Conservation: A Reference for Managing Wild Animal and Plant Populations*, edited by C. M. Schonewald-Cox, S. M. Chambers, B. MacBryde, and W. L. Thomas. Benjamin/Cummings, Menlo Park, CA.

Rebholz, W., and E. Harley. 1999. Phylogenetic relationships in the bovid subfamily Antilopinae based on mitochondrial DNA sequences. *Molecular Phylogenetics and Evolution* 12(2):87–94.

Rietkerk, F., and A. Glatston. 2003. *European Collection Plan for Antelopes.* EAZA Antelope and Giraffe TAG. 31 pp.

Roldan, E. R. S., J. Cassinello, T. Abáigar, and M. Gomendio. 1998. Inbreeding, fluctuating asymmetry, and ejaculate quality in an endangered ungulate. *Proceedings of the Royal Society B–Biological Sciences* 265:243–48.

Roney, E. E., Jr. 1978. San Antonio Zoo antelope collection. In *AAZPA Regional Workshop Proceedings 1977–78.* Hill's Division of Riviana Foods, Topeka, KS. Pp. 100–109.

Rothschild, W. 1921. Captain Angus Buchanan's Air Expedition, III. Ungulate mammals collected by Captain Angus Buchanan. *Novitates Zoologicae* 28:75–77.

Ruiz-López, M. J., N. Gañan, J. A. Godoy, A. Del Olmo, J. Garde, G. Espeso, A. Vargas, F. Martinez, E. R. S. Roldán, and M. Gomendio. 2012. Heterozygosity-fitness correlations and inbreeding depression in two critically endangered mammals. *Conservation Biology* 26:1121–29. doi: 10.1111/j.1523–1739.2012.01916.x.

Ruiz-López, M. J., E. R. S. Roldán, G. Espeso, and M. Gomendio. 2009. Pedigrees and microsatellites among endangered ungulates: What do they tell us? *Molecular Ecology* 18:1352–64.

RZSS and IUCN Antelope Specialist Group. 2014. *Conservation Review of the Dama Gazelle* (Nanger dama). Royal Zoological Society of Scotland, Edinburgh, United Kingdom. 70 pp.

Saleh, M. A. 2001. Egypt. Chap. 7 in *Antelopes: Global Survey and Regional Action Plans, Part 4. North Africa, the Middle East, and Asia*, compiled by D. P. Mallon and S. C. Kingswood. SSC Antelope Specialist Group, IUCN, Gland, Switzerland. Pp. 48–54.

Sayer, J. A. 1977. Conservation of large mammals in the republic of Mali. *Biological Conservation* 12:245–63.

Scholte, P. 2013. *Nanger dama* dama gazelle. In *Mammals of Africa*, vol. 1, *Pigs, Hippotamuses, Chevrotain, Giraffes, Deer and Bovids*, edited by J. Kingdon and M. Hoffmann. Bloomsbury, London. Pp. 382–87.

Sclater, P. L., and O. Thomas. 1897–1898. *The Book of Antelopes.* Vol. 3. R. H. Porter, London, United Kingdom. 245 pp.

Seaman, D. E., and R. A. Powell. 1996. An evaluation of the accuracy of kernel density estimators for home range analysis. *Ecology* 77:2075–85.

Senn, H., L. Banfield, T. Wacher, J. Newby, T. Rabeil, J. Kaden, A. C. Kitchener, T. Abáigar, T. Luísa Silva, M. Maunder, and R. Ogren. 2014. Splitting or lumping? A conservation dilemma exemplified by the critically endangered dama gazelle (*Nanger dama*). *PLoS ONE* 9(6):e98693. doi:10.1371/journal.pone.0098693.

Senn, H., T. Wacher, J. Newby, A. Matchano, E. C. Mungall, B. Pukazhenthi, A. Kitchener, A. Eyres, and T. Rabeil. 2016. Update: Genetic relatedness of critically endangered dama gazelle populations in the wild and captivity. *Gnusletter* 33(1):5–8.

Sgrò, C. M., A. J. Lowe, and A. A. Hoffmann. 2011. Building evolutionary resilience for conserving biodiversity under climate change. *Evolutionary Applications* 4:326–37.

Siegel, M. 2013. *Historical Progeny Sex Ratio at White Oak.* Unpublished technical report for White Oak Conservation Holdings, Yulee, FL. 3 pp.

Silver, H. 1961. Deer milk compared with substitute milk for fawns. *Journal of Wildlife Management* 25(1):66–70.

Speeg, B., S. Shurter, and A. Eyres. 2014. Dama gazelle (*Nanger dama*) breeding programs at two conservation centers in the U.S.A. *Gnusletter* 31(1):5–9.

Spottiswoode, C., and A. P. Møller. 2004. Genetic similarity and hatching success in birds. *Proceedings, Biological Sciences, the Royal Society* 271:267–72.
Strandberg, Å. 2009. The Gazelle in Ancient Egyptian Art: Image and Meaning. PhD diss., Uppsala Universitet, Uppsala, Sweden. 262 pp.
Swindell, W. R., and J. L. Bouzat. 2005. Modeling the adaptive potential of isolated populations: Experimental simulations using *Drosophila*. *Evolution* 59:2159–69.
Their, T. 2015. *AZA North American Regional Studbook for Addra Gazelle.* Saint Louis Zoo, Saint Louis, MO.
Thomassey, J. P., and J. E. Newby. 1990. Chad. Chap. 6 in *Antelopes: Global Survey and Regional Action Plans, Part 3, West and Central Africa*, edited by R. East. IUCN, Gland, Switzerland. Pp. 22–28.
Thompson, M. V. 2006. *President George Washington's Deer.* Research summary provided to Marida Favia del Core Borromeo by Mary V. Thompson, research specialist, Mount Vernon Estate and Gardens, January 20.
Traweek, M. S. 1985. *Statewide Census of Exotic Big Game Animals.* Job no. 21, federal aid project no. W-109-R-3, job performance report. Texas Parks and Wildlife Department, Austin. 33 pp.
———. 1989. *Statewide Census of Exotic Big Game Animals.* Job no. 21, federal aid project no. W-109-R-12, job performance report. Texas Parks and Wildlife Department, Austin. 52 pp.
———. 1995. *Statewide Census of Exotic Big Game Animals.* Job no. 21, federal aid project no. W-127-R-3, job performance report. Texas Parks and Wildlife Department, Austin. 53 pp.
Treus, V., and D. Kravchenko. 1968. Methods of rearing and economic utilization of eland in the Askaniya-Nova Zoological Park. In *Comparative Nutrition of Wild Animals*, edited by M. A. Crawford. Symposium of the Zoological Society of London, no. 21. Pp. 395–411.
Trotignon, J. 1975. *Le statut et la conservation de l'addax et de l'oryx et de la faune associée en Mauritanie.* Report to IUCN. Morges, Switzerland. 36 pp.
Valverde, J. A. 1957. *Aves del Sahara espagnol (Estudio ecológico del desierto).* Consejo Superior de Investigaciones Científicas, Instituto de Estudios Africanos, Madrid.
———. 2004. Sáhara, Guinea, Marruecos, Expediciones africanas. *Memorias de un Biólogo Heterodoxo*, Tomo III. Editorial Quercus, Consejo Superior de Investigaciones Científicas, Madrid, Spain. 272 pp.
van Zyl, J. H. M., and A. S. Wehmeyer. 1970. The composition of the milk of springbok (*Antidorcas marsupialis*), eland (*Taurotragus oryx*), and black wildebeest (*Connochaetus* [sic] *gnou*). *Zoologica Africana* 5(1):131–33.
Vassart, M., A. Greth, V. Durand, and E. Cribiu. 1993. An unusual *Gazella dama* karyotype. *Annales de Génétique* 36:111–17.
Vassart, M., A. Séguéla, and H. Hayes. 1994. Chromosomal evolution in gazelles. *Journal of Heredity* 86(3):216–27.
Vice, T. E., and F. H. Olin. 1967. A note on the milk analysis and hand-rearing of the greater kudu *Tragelaphus strepsiceros* at San Antonio Zoo. *International Zoo Yearbook* 9(1):114.
Vilas, C., E. S. A. N. Miguel, R. Amaro, and C. Garcia. 2006. Relative contribution of inbreeding depression and eroded adaptive diversity to extinction risk in small populations of shore campion. *Conservation Biology* 20:229–38.
Vincke, P. P. 1987. Plateau d'El Aguer, compte rendu de mission. In *Pour une gestion de la faune du Sahel*, edited by P. P. Vincke, G. Sournia, and E. Wangari. *Environnement Africain, Série Etudes et Recherches*, no. 120–21. MAB/ENDA/IUCN. Pp. 104–10.
Wacher, T., and J. Newby. 2010. *Wildlife and Land Use Survey of the Manga and Eguey Regions, Chad.* Pan Saharan Wildlife Survey, technical report no. 4, August. Sahara Conservation Fund, L'Isle, Switzerland. 70 pp.
Wacher, T., J. Newby, and M. H. Hacha. 2014. Conservation of the dama gazelle. *Survey of the Manga & Western Chad, January–February 2014.* Sahara Conservation Fund and Zoological Society of London. 50 pp.
Wacher, T., J. E. Newby, S. T. Monfort, J. Tubiana, D. Moksia, W. Houston, and A. M. Dixon. 2004. Sahelo-Saharan Interest Group antelope update, Chad 2001 and Niger 2002. *Antelope Survey Update* no. 9. IUCN/SSC Antelope Specialist Group Report. Pp. 52–59.
Wacher, T., D. Potgieter, M. H. Hacha, S. Dogringar, and T. Rabeil. 2015. *Dama Gazelle Survey, the Manga Region, Western Chad, February 2015.* Zoological Society of London, African Parks Network, and Sahara Conservation Fund, L'Isle, Switzerland. 28 pp.
Walther, F. R. 1972a. Subfamily: Antilopinae. In *Grzimek's Animal Life Encyclopedia*, edited by B. Grzimek. Van Nostrand Reinhold, New York. Pp. 431–49.
———. 1972b. Territorial behaviour in certain horned ungulates, with special reference to the examples of Thomson's and Grant's gazelles. *Zoologica Africana* 7(1):303–7.
———. 1984. *Communication and Expression in Hoofed Mammals.* Indiana University Press, Bloomington. 423 pp.

Walther, F. R., E. C. Mungall, and G. A. Grau. 1983. *Gazelles and Their Relatives: A Study in Territorial Behavior.* Noyes, Park Ridge, NJ. 239 pp.

Weeks, A. R., C. M. Sgro, A. G. Young, R. Frankham, N. J. Mitchell, K. A. Miller, M. Byrne, D. J. Coates, M. D. B. Eldridge, P. Sunnucks, M. F. Breed, E. J. James, and A. A. Hoffmann. 2011. Assessing the benefits and risks of translocations in changing environments: A genetic perspective. *Evolutionary Applications* 4:709–25. doi: 10.1111/j.1752-4571.2011.00192.x.

Wiesner, H., and P. Müller. 1998. On the reintroduction of the mhorr gazelle in Tunisia and Morocco. *Naturwissenschaften* 85:553–55.

Wilson, R. T. 1980. Wildlife in northern Darfur, Sudan: A review of its distribution and status in the recent past and at present. *Biological Conservation* 17:85–101.

Wojtusik, J., J. L. Brown, and B. S. Pukazhenthi. 2017. Non-invasive hormonal characterization of the ovarian cycle, pregnancy, and seasonal anestrus of the female addra gazelle (*Nanger dama ruficollis*). *Theriogenology* 95:96–104.

Woodfine, T., T. Gilbert, and H. Engel. 2004. A summary of past and present initiatives for the conservation and reintroduction of addax and scimitar-horned oryx in North Africa. In *Proceedings of the EAZA Conference 2004*, Kolmarden, 21–25 September. Pp. 208–11.

Wright, P. A., E. F. Deysher, and C. A. Cary. 1939. Variations in the composition of milk. In *Food and Life: Yearbook of Agriculture 1939.* USDA, Washington, DC. Pp. 639–48.

Zboray, A. 2009. *Rock Art of the Libyan Desert.* 2nd ed. (DVD ROM). Fliegel Jezerniczky Ltd., Newbury, United Kingdom.

Zerbe, P., M. Clauss, D. Codron, L. B. Lackey, E. Rensch, J. W. Streich, J-M. Hatt, and D. W. Müller. 2012. Reproductive seasonality in captive wild ruminants: Implications for biogeographical adaptation, photoperiodic control, and life history. *Biological Reviews* 87(4):965–90.

Contributors

Teresa Abáigar, research scientist, Estación Experimental de Zonas Áridas (Spanish National Research Council, CSIC), Almería, Spain

Hessa Al Qahtani, Unit Head Conservation Programmes, Al Ain Zoo, United Arab Emirates

Lisa Banfield, Unit Head Conservation Research, Al Ain Zoo, United Arab Emirates

Mark Craig, director of life sciences, Al Ain Zoo, United Arab Emirates

Gerardo Espeso Pajares, EEP coordinator for *Gazella dama mhorr*, Estación Experimental de Zonas Áridas, Finca Experimental La Hoya, Almería, Spain

Adam Eyres, hoofstock curator, Fossil Rim Wildlife Center, Glen Rose, Texas, USA

Tania Gilbert, conservation biologist, Marwell Wildlife, Winchester, United Kingdom

Abdelkader Jebali, vice-president of the Tunisia Wildlife Conservation Society, Tunisia

Andrew C. Kitchener, principal curator of vertebrates, National Museums Scotland, Edinburgh, Scotland, United Kingdom

David Mallon, cochair of the IUCN Species Survival Commission's Antelope Specialist Group, United Kingdom

Elizabeth Cary Mungall, science officer, Second Ark Foundation, Ingram, Texas, USA, and adjunct professor, Department of Biology, Texas Woman's University, Denton, Texas, USA

John Newby, chief executive officer, Sahara Conservation Fund, Switzerland

Thomas Rabeil, regional program officer, Sahara Conservation Fund, France

Helen Senn, research scientist (conservation genetics), WildGenes Laboratory, Royal Zoological Society of Scotland, Edinburgh Zoo, Edinburgh, Scotland, United Kingdom

Tim Wacher, senior wildlife biologist, Zoological Society of London, United Kingdom

Frans M. van den Brink, animal dealer, De Krim, the Netherlands

Bonnie C. Yates, morphology section chief (retired), US Fish and Wildlife Service National Fish and Wildlife Forensics Laboratory, Ashland, Oregon, USA

Index